Bilingual edition

計測工学

高　偉
清水裕樹
羽根一博
祖山　均
足立幸志［著］

Measurement and Instrumentation

Wei Gao
Yuki Shimizu
Kazuhiro Hane
Hitoshi Soyama
Koshi Adachi

朝倉書店

Bilingual edition

Keisoku Kogaku
Measurement and Instrumentation

Wei Gao, Yuki Shimizu, Kazuhiro Hane, Hitoshi Soyama, Koshi Adachi

ISBN 978-4-254-20165-9

Published by Asakura Publishing Company, Ltd.

まえがき

近年，気候変動，エネルギー危機など地球規模の課題が増加する中，国際的な科学技術協力の必要性が高まっている．また，産業の国際化に伴い，日本企業にとっては，国内に立脚しながらグローバルな視野に立つ事業展開が不可欠である．それらに対応するためには，高度な専門知識を日本語のみならず「世界の公用語」といわれる英語でも活用できる能力を備えた研究者，技術者の育成が求められる．

一方，留学生が急速に増える中，国内の大学では英語による授業などの実施体制の構築が行われている．著者らが所属する東北大学工学部でも英語による授業のみで学位が取得できる国際学士コースが開設された．基本的にこれまでの日本語コースと同一のカリキュラムになっているため，同じ授業が日本語と英語のクラスで並列開講されている．このような背景の中，国内外初の試みとなるが，「計測工学」を日本語と英語の両方で学べる教科書として，講義を担当している著者らが協力して本書を執筆した．

本書の最大の特徴は，計測工学の基礎から最新技術までの重要事項を精選し，日本語と英語で分かりやすく記述していることである．計測工学は極めて範囲の広い学問分野であり，これまでの教科書は計測工学全般を浅く広く羅列する便覧的なものが多かったが，本書ではあえて重要な内容に焦点を絞って，その基本となる数学的・物理的原理を丁寧に記述している．また，図，表および式がページの中央に，上下にはそれぞれ和文と英文の記述が対応するように配置し，効率的かつ明快なページレイアウトにしている．

本書が，日本人学生と技術者が計測工学を学ぶ際，英語での学習をサポートし，また，計測工学を学ぶ留学生や国内企業で働く外国人技術者にとっても日本語に触れる機会を提供することのできる一冊となれば幸いである．

本書の出版にあたり，朝倉書店編集部に多大な御尽力を賜ったこと，また東北大学教員の伊東聡先生，陳遠流先生，学生の牛増渊君，齋藤俊樹君らに原稿チェックや図面作成などの御協力を頂いたことを記し，心から感謝申し上げる．

2017 年 2 月

著者代表　高　偉

Preface

In recent years, there has been an increasing need for international cooperation of science and technology in responding to the growing numbers of global problems of climate change, energy crisis, and so on. The rapid globalization of industries also requires Japanese companies to expand the business domestically and internationally. It is thus necessary for Japanese researchers and engineers to have a high level of expertise in both Japanese and English.

Meanwhile, English-taught degree programs are being established in Japanese universities for international students. In Tohoku University, an international course (IMAC) has been launched in the School of Engineering. The curriculum of IMAC is basically the same as that of the regular Japanese course and the same subjects are taught in Japanese and English classes in parallel. Under such background, this book, as the first Japanese-English bilingual textbook on measurement and instrumentation, has been written by the authors who are teaching the parallel classes.

The best feature of this book is that important points of measurement and instrumentation, from fundamental theories to the newest technologies, are clearly explained in both Japanese and English. Measurement and instrumentation is multidisciplinary in essence and covers broad areas. For this reason, most of the existing textbooks are written in a handbook style with brief descriptions on wide range of topics. In contrast, this book is focused on essential elements with detailed explanations on the mathematical and physical principles. In addition, an efficient and easy-to-read page layout is realized by locating the figures, tables and equations in the middle of the page put between Japanese and English texts.

This book can be used as a textbook or a reference book by either Japanese or English speaking students and engineers for learning measurement and instrumentation in their own languages. It also provides the opportunity to learn the other language on the same subject.

We would like to thank Asakura Publishing for their dedicated efforts on editing of this book. The assistance from Prof. So Ito, Prof. Yuan-Liu Chen, Mr. Zengyuan Niu, Mr. Toshiki Saito of Tohoku University is also appreciated.

February, 2017

Wei Gao, on behalf of the authors

目　次
Table of Contents

第1章　計測の基準

計測とは，何らかの目的を持って，事物を量的にとらえるための方法・手段を考究し，実施し，その結果を用いて目的を達成させること，である．ここでいう事物を量的にとらえることは対象となる物理量を定量化する意味である．それには何らかの基準が必要である．本章では計測の概念を述べた後，計測の基準を中心に記述する．

● 1.1　計測の概念

計測という行為には，

・計測の目的および定量化しようとする物理量を明確にすること

・対象の物理量を定量化するために計測機を選択あるいは開発し，測定を行うこと

・測定結果を目的達成のために有効利用すること

という手順と要素が含まれる．

Chapter 1　Standards of Measurement

Measurement and instrumentation, often simply called measurement, is the planning and implementation of methods and means to quantify a thing, and then utilizing the result for a certain purpose. Here "quantify a thing" means quantification based on a certain standard. The concept and standards of measurement are described in this chapter.

● 1.1　Concept of measurement

The action of measurement contains the following procedures and operations:

· to make clear the objective and the physical quantity to be measured

· to select or develop an instrument and to make the measurement for quantification of the physical quantity

· to utilize the measurement result for achieving the objective

One purpose of measurement is for control of a machine or a system. In the

計測の目的はまず機械や装置の制御に用いられることが考えられる．衝突被害軽減ブレーキ機能あるいは自動ブレーキ機能を搭載している自動車の場合には，衝突被害を軽減させる目的で，障害物までの距離（計測量）を自動車に搭載したレーダやカメラなどのセンサ（計測機）で測定し，その測定結果をブレーキの制御装置にフィードバックし，自動ブレーキの操作に利用する，という計測の行為が行われている．計測のもう一つの重要な目的は，製造工程における品質管理である．例えば半導体製造プロセスにおける前工程の最終段階では，良品・不良品を仕分ける目的で，ウエハをチップごとに測定をして，その結果を利用して良品だけに後工程を施す．また，最終製品の出荷検査のために再度測定を行い，不良品を除いた形で製品を出荷する．

一方，科学研究の分野では新しい事実や発見のため，また技術開発の分野では新しい技術や製品の特性評価のためにも計測がよく行われている．例えばブラックホールの解明などをめざしてKAGRA計画（大型低温重力波望遠鏡計画）が進められている．重力は，宇宙の構造や進化を支配する重要なものであり，宇宙の謎を解き明かす重要な鍵となっている．この計画では，重力波を観測する目的で，レーザ干渉計を地底深くに設置し，重力波によってゆがんだ空間において直交方向に走る2本のレーザ光の光路差を測定する．その測定結果を用いて重力がもとになって生まれる宇宙から

automatic braking system of an automobile, a sensor is employed to measure the distance between the automobile and another vehicle, person, or obstacle, and the measurement result is utilized as a feedback signal by the braking system for controlling the automobile to avoid an imminent collision without any driver input. Another purpose of measurement and the related instrumentation is for quality control in a manufacturing process. For example, for the purpose of identifying defective chips, each chip on the circuit board is measured in the final stage of the pre-process of semiconductor manufacturing so that only the non-defective chips will be sent to the post-process. The chips are then measured again for assurance that only non-defective chips will be shipped.

Measurement is also essential for investigating new facts and findings in scientific research and for evaluating new technologies and new products in technology development. For example, a scientific project called KAGRA (Kamioka Large-scale Cryogenic Gravitational Wave Telescope) investigates black holes in the cosmos through direct detection of gravitational waves. For this purpose, a laser interferometer is employed to detect the optical path difference between two laser

の波動「重力波」をとらえている．

いうまでもなく，測定は物理量の計測という行為にとって最も中心となる要素である．計測対象となる量を，基準として用いる量と比較し，数値または符号を用いて表すことは測定の定義となっている．すなわち，対象となる物理量（計測量）を何らかの方法で基準となる同種の量と比較して，計測量の大きさが基準量の大きさの何倍であるかを定めることによって，計測量を定量化させる．計測量 Q の大きさを $|Q|$，基準量 u の大きさを $|u|$ とし，$|Q|$ と $|u|$ を比較した結果，$|Q|$ が $|u|$ の n 倍であることが分かったとすると，計測量 Q を式(1-1)のように定量化，つまりその大きさを数値で表すことができる．物理量（計測量）の大きさを数値で表すための基準量の大きさのことを単位と呼んでいる．

● 1.2　単位と単位系

物理学における一定の体系の下で物理量の性質を表す次元が確定し，定められた単位の倍数としてその次元の量の大きさを表すことができる．単位は物理量を定量化するための基準であり，それには明確で使いやすい定義が必要である．また，物理量はたくさん存在するので，それらに対応する単位を体系化することが実用上重要であ

$$|Q| = n\,|u| \tag{1-1}$$

beams, from which the gravitational waves can be detected.

The target quantity, often called a measurand, is quantified by the measurement operation. That operation is the process of ascertaining the value of a measurand through comparison with a standard of the quantity so that the value of measurand can be quantified as a simple multiple of the standard quantity. Assume the values of the measurand Q and the standard quantity u are expressed by $|Q|$ and $|u|$ respectively. If $|Q|$ is identified as n times $|u|$ as the result of the comparison, the measurand Q is quantified by Eq. (1-1) and the value of Q can be expressed in numerical form. The standard quantity used in the measurement is referred to as the measurement unit.

● 1.2　Measurement units and the system of units

The properties of a physical quantity are expressed by dimensions that are determined in physics. The value of the dimension of the quantity, often simply referred to as the value of the quantity, is expressed by a simple multiple of the measurement unit. The definition of a unit must be clear and easy to use. Because of

る．単位を体系化したものを単位系という．単位系では，いくつかの量を選び，その他のすべての量を物理方程式に基づいてこれらの基本量の組合せで表すことになっている．最初に選んだ量を基本量，その単位を基本単位といい，その他の量を組立量，組立量の単位を組立単位という．単位系の構築において，基本単位の選択は最も基本的で重要なものである．基本単位の選定条件として，

・基本的であること

・その次元が互いに独立であること

・数がなるべく少ないこと

が挙げられる．

現在国際的に唯一認められているのは国際単位系 (SI) である．SI 単位系は7つの基本量／基本単位，基本量／基本単位を組み合わせた組立量／組立単位，および20の接頭語からなる．

Table 1-1 に SI 基本単位，基本量およびそれに対応する次元の記号を示す．SI 単位系の基本量／基本単位は，長さ／メートル，質量／キログラム，時間／秒，電流／アンペア，熱力学温度／ケルビン，物質量／モル，光度／カンデラとなっている．ま

the large number of physical quantities, it is also important to treat the different units as a system for practical applications. In a system of units, a small number of quantities are selected to express other quantities based on physics equations. The selected quantities and their units are called base quantities and base units, respectively. Other quantities and units are called derived quantities and derived units respectively. Selection of the base quantities/units is the most fundamental and important step for the establishment of a system of units. Some conditions for this selection are:

・The base quantities must be fundamental quantities in physics.

・The dimensions of the base quantities must be independent from one another.

・The number of base quantities must be kept to a minimum.

The International System of Units (SI) is the only system of units that is recognized around the world. It is composed of seven base quantities/units, derived quantities/units, and 20 prefixes.

The seven base units of the SI are listed in Table 1-1, in which each base quantity is related to a unit name and a unit symbol. The base quantities/units are length/meter, mass/kilogram, time/second, electric current/ampere, thermodynamic

た，それぞれの基本単位は以下のように定義される．

■秒：単位 Hz（s^{-1}に等しい）による表現で，基底状態でのセシウム 133 原子の超微細構造の遷移周波数 $\Delta\nu_{Cs}$ の数値を 9 192 631 770 と定めることによって定義される．

■メートル：単位 m s^{-1} による表現で，真空中の光の速さ c の数値を 299 792 458 と定めることによって定義される．

■キログラム：単位 J s（kg m^2 s^{-1} に等しい）による表現で，プランク定数 h の数値を 6.62607015×10^{-34} と定めることによって定義される．

■アンペア：単位 C (A s に等しい) による表現で，電気素量 e の数値を 1.602176634

Table 1-1　SI base units and base quantities

Base quantity 基本量		SI base unit SI 基本単位	
Name 名称	Symbol (dimension symbol) 記号 (次元記号)	Name 名称	Symbol 記号
length 長さ	l, x, r, etc. (L)	meter メートル	m
mass 質量	m(M)	kilogram キログラム	kg
time 時間	t(T)	second 秒	s
electric current 電流	I, i(I)	ampere アンペア	A
thermodynamic temperature 熱力学温度	T(Θ)	kelvin ケルビン	K
amount of substance 物質量	n(N)	mole モル	mol
luminous intensity 光度	I_v(J)	candela カンデラ	cd

temperature/kelvin, amount of substance/mole, and luminous intensity/candela. The base units are defined as follows:

■ The second is defined by taking the fixed numerical value of the caesium frequency $\Delta\nu_{Cs}$, the unperturbed ground-state hyperfine transition frequency of the caesium 133 atom, to be 9 192 631 770 when expressed in the unit Hz, which is equal to s^{-1}.

■ The meter is defined by taking the fixed numerical value of the speed of light in vacuum c to be 299 792 458 when expressed in the unit m s^{-1}.

■ The kilogram is defined by taking the fixed numerical value of the Planck constant h to be 6.62607015×10^{-34} when expressed in the unit J s, which is equal to kg $m^2\cdot s^{-1}$.

$\times 10^{-19}$ と定めることによって定義される．

■ケルビン：単位 $\mathrm{J\,K^{-1}}$（$\mathrm{kg\,m^2\,s^{-2}\,K^{-1}}$ に等しい）による表現で，ボルツマン定数 k の数値を 1.380649×10^{-23} と定めることによって定義される．

■モル：1モルは正確に 6.02214076×10^{23} の要素粒子を含む．この数値は単位 $\mathrm{mol^{-1}}$ による表現で，アボガドロ定数 N_{A} の固定された数値であり，アボガドロ数と呼ばれる．

■カンデラ：単位 $\mathrm{lm\,W^{-1}}$（$\mathrm{cd\,sr\,W^{-1}}$ あるいは $\mathrm{cd\,sr\,kg^{-1}\,m^{-2}\,s^3}$ に等しい）による表現で，周波数 540×10^{12} Hz の単色光の発光効率 K_{cd} の数値を 683 と定めることによって定義される．

単位記号と次元記号はローマン体（立体）を用いるが，量記号は一般にイタリック体（斜体）で表される．一方，SI 単位系における組立量 Q の次元 $\dim Q$ は，組立量と SI 基本量の関係式に従って，式(1-2)のように基本量の次元のべき乗の積で表される．式にある L, M, T, I, Θ, N, J はそれぞれ7つの SI 基本量の次元である．指数 $\alpha, \beta, \gamma, \delta, \varepsilon, \zeta, \eta$ は，正か負かゼロである小さい整数で，次元指数と呼ばれる．組立量 Q の単位は，式(1-2)と同様な関係式で SI 基本量のべき乗の積で表される．Table 1-2 に組立量と組立単位の例を示す．同じ種類の量の比で定義される物理量は次元指数がすべてゼロとなる．そのような量は無次元の量と呼ばれ，その組立単位は1であり，単位記号は書かないことになっている．Table 1-2 にある屈折率と比透磁率は無次元

$$\dim Q = \mathrm{L}^{\alpha}\mathrm{M}^{\beta}\mathrm{T}^{\gamma}\mathrm{I}^{\delta}\Theta^{\varepsilon}\mathrm{N}^{\zeta}\mathrm{J}^{\eta} \qquad (1\text{-}2)$$

■ The ampere is defined by taking the fixed numerical value of the elementary charge e to be $1.602176634\times 10^{-19}$ when expressed in the unit C, which is equal to A s.

■ The kelvin is defined by taking the fixed numerical value of the Boltzmann constant k to be 1.380649×10^{-23} when expressed in the unit $\mathrm{J\,K^{-1}}$, which is equal to $\mathrm{kg\,m^2\,s^{-2}\,K^{-1}}$.

■ One mole contains exactly 6.02214076×10^{23} elementary entities. This number is the fixed numerical value of the Avogadro constant, N_{A}, when expressed in the unit $\mathrm{mol^{-1}}$ and is called the Avogadro number.

■ The candela is defined by taking the fixed numerical value of the luminous efficacy of monochromatic radiation of frequency 540×10^{12} Hz, K_{cd}, to be 683 when expressed in the unit $\mathrm{lm\,W^{-1}}$, which is equal to $\mathrm{cd\,sr\,W^{-1}}$, or $\mathrm{cd\,sr\,kg^{-1}\,m^{-2}\,s^3}$.

Unit names and dimension symbols are normally written in roman (upright) type while symbols for quantities are set in an italic font. As shown in Eq. (1-2), the dimension of a derived quantity is written as the products of powers of the dimensions of the base quantities using the equations that relate the derived quantities to the base

の量の例である．屈折率は真空中の光の速さと媒質中の光の速さの比であり，比透磁率は媒質の透磁率と真空の透磁率との比である．

使用上の便利さを考慮して，SI は Table 1-3 に示す 22 の組立単位に対してその単

Table 1-2 Examples of derived units and derived quantities

Derived quantity 組立量		SI derived unit SI 組立単位	
Name 名称	Symbol 記号	Name 名称	Symbol 記号
area 面積	A	square meter 平方メートル	m^2
volume 体積	V	cubic meter 立方メートル	m^3
speed, velocity 速度	v	meter per second メートル毎秒	m/s
acceleration 加速度	a	meter per second squared メートル毎秒毎秒	m/s^2
wavenumber 波数	$\sigma, \tilde{\nu}$	reciprocal meter 毎メートル	m^{-1}
density, mass density 質量密度	ρ	kilogram per cubic meter キログラム毎立方メートル	kg/m^3
surface density 面積密度	ρ_A	kilogram per square meter キログラム毎平方メートル	kg/m^2
specific volume 比密度	ν	cubic meter per kilogram 立方メートル毎キログラム	m^3/kg
current density 電流密度	j	ampere per square meter アンペア毎平方メートル	A/m^2
magnetic field strength 磁界の強さ	H	ampere per meter アンペア毎メートル	A/m
amount concentration, concentration 濃度	c	mole per cubic meter モル毎立方メートル	mol/m^3
mass concentration 質量濃度	ρ, γ	kilogram per cubic meter キログラム毎立方メートル	kg/m^3
luminance 輝度	L_v	candela per square meter カンデラ毎平方メートル	cd/m^2
refractive index 屈折率	n	one （数字の）1	1
relative permeability 比透磁率	μ_r	one （数字の）1	1

quantities indicated by L, M, T, I, Θ, N, and J. The exponents $\alpha, \beta, \gamma, \delta, \varepsilon, \zeta$, and η are called the dimensional exponents, which are generally small integers which can be positive, negative or zero. The unit of the derived quantity Q is also written as the products of powers of the base units in the form of Eq. (1-2). Table 1-2 shows the

Table 1-3 Derived units in the SI with special names and symbols

Derived quantity 組立量	Name 名称	Symbol 記号	Expressed in terms of other SI units 他の SI 単位による表し方
plane angle 平面角	radian ラジアン	rad	m/m=1
solid angle 立体角	steradian ステラジアン	sr	$m^2/m^2=1$
frequency 周波数	hertz ヘルツ	Hz	s^{-1}
force 力	newton ニュートン	N	$m\ kg\ s^{-2}$
pressure, stress 圧力，応力	pascal パスカル	Pa	$N/m^2=m^{-1}\ kg\ s^{-2}$
energy, work, amount of heat エネルギー，仕事，熱量	joule ジュール	J	$N\ m=m^2\ kg\ s^{-2}$
power, radiant flux 仕事率，放射束	watt ワット	W	$J/s=m^2\ kg\ s^{-3}$
electric charge, amount of electricity 電荷，電気量	coulomb クーロン	C	s A
electric potential difference, electromotive force 電位差，起電力	volt ボルト	V	$W/A=m^2\ kg\ s^{-3}\ A^{-1}$
capacitance 静電容量	farad ファラド	F	$C/V=m^{-2}\ kg^{-1}\ s^4\ A^2$
electric resistance 電気抵抗	ohm オーム	Ω	$V/A=m^2\ kg\ s^{-3}\ A^{-2}$
electric conductance コンダクタンス	siemens ジーメンス	S	$A/V=m^{-2}\ kg^{-1}\ s^3\ A^2$
magnetic flux 磁束	weber ウェーバ	Wb	$V\ s=m^2\ kg\ s^{-2}\ A^{-1}$
magnetic flux density 磁束密度	tesla テスラ	T	$Wb/m^2=kg\ s^{-2}\ A^{-1}$
inductance インダクタンス	henry ヘンリー	H	$Wb/A=m^2\ kg\ s^{-2}\ A^{-2}$
Celsius temperature セルシウス温度	degree Celsius セルシウス度	℃	K
luminous flux 光束	lumen ルーメン	lm	cd sr=cd
illuminance 照度	lux ルクス	lx	$lm/m^2=m^{-2}\ cd$
activity referred to a radionuclide 放射性核種の放射能	becquerel ベクレル	Bq	s^{-1}
absorbed dose, specific energy (imparted), kerma 吸収線量，比エネルギー分与，カーマ	gray グレイ	Gy	$J/kg=m^2\ s^{-2}$
dose equivalent, ambient dose equivalent 線量当量，周辺線量当量	sievert シーベルト	Sv	$J/kg=m^2\ s^{-2}$
catalytic activity 酵素活性	katal カタール	kat	$s^{-1}\ mol$

位の固有の名称と単位記号を認めている．これらは使用頻度が高い量の単位で基本単位の組合せによる表現よりはるかに簡単になる．例えば熱量の組立単位 $m^2\,kg\,s^{-2}$ は1つの単位記号Jで記述できるようになっている．Table 1-3 にあるヘルツとベクレルという固有名称を持つ単位はともにSI基本単位“秒”の逆数 s^{-1} で表されるが，それぞれ異なる物理量を表している．また，いかなる量もただ1つのSI単位を持つが，Table 1-3 に示す固有の単位記号を使うと幾通りかの表現があり得る．

単独のSI単位の大きさよりはるかに大きい量あるいは小さい量を表す際に，SI単位と併用される一組20個の接頭語が決められている．Table 1-4 にSI単位系の接頭

Table 1-4　SI prefixes

Factor 乗数	Name 名称	Symbol 記号	Factor 乗数	Name 名称	Symbol 記号
10^{1}	deka デカ	da	10^{-1}	deci デシ	d
10^{2}	hecto ヘクト	h	10^{-2}	centi センチ	c
10^{3}	kilo キロ	k	10^{-3}	milli ミリ	m
10^{6}	mega メガ	M	10^{-6}	micro マイクロ	μ
10^{9}	giga ギガ	G	10^{-9}	nano ナノ	n
10^{12}	tera テラ	T	10^{-12}	pico ピコ	p
10^{15}	peta ペタ	P	10^{-15}	femto フェムト	f
10^{18}	exa エクサ	E	10^{-18}	atto アト	a
10^{21}	zetta ゼタ	Z	10^{-21}	zepto ゼプト	z
10^{24}	yotta ヨタ	Y	10^{-24}	yocto ヨクト	y

examples of derived quantities/units.

For a derived quantity Q that is defined as the ratio of two quantities of the same kind, all of the dimensional exponents in the expression for the dimension of Q are zero. Such a quantity is described as being dimensionless. For example, the dimensionless refractive index in Table 1-2 is defined as the ratio of the speed of light in vacuum to that in the medium. Similarly, the dimensionless relative permeability is the ratio of the permeability in the medium to that in vacuum.

For convenience, certain derived units have been given special names and symbols, as listed in Table 1-3. The special names and symbols are simply a compact form of the expression of combinations of base units that are used frequently. For example, the derived unit $m^2\,kg\,s^{-2}$ for the amount of heat is expressed simply by J. The hertz and becquerel in Table 1-3 are for different quantities, although both are expressed by s^{-1}. It should be noted that each quantity has a unique SI unit but can have several different expressions, using the special symbols shown in Table 1-3.

A series of twenty prefixes are approved in SI units for the convenience of

語を示す．接頭語を利用すると，表現上便利になる．接頭語の記号はローマン体で表す．

歴史的な事情により，質量の基本単位キログラムにははじめからキロという接頭語がついている．キログラムの10進の倍量および分量を表す場合，接頭語をつけるときには，キログラムに接頭語をつけずに，グラム（g）にただ1つだけ接頭語をつける．例えば，マイクロキログラム（μkg）とはせず，ミリグラム（mg）とする．

● 1.3 トレーサビリティと計測標準

計測機を使って正確に計測を行うためには，その計測機はより正確な（不確かさがより小さい）標準器によって校正されることが必要条件である．この標準器もより正確な標準器によって校正される，というようにより正確な標準器をもとめていくと国家標準あるいは国際標準に辿り着く．計測機が校正の連鎖によって国家標準あるいは国際標準に辿り着けることが確かめられている場合，この計測機により得られた結果は国家標準にトレーサブルであるという．トレーサビリティとは，「不確かさがすべて表記された切れ目のない比較の連鎖によって，決められた基準に結びつけられ得る測定結果または標準の値の性質である．基準は通常，国家標準または，国際標準」と

expressing the values of quantities that are much larger than or much smaller than the SI unit. Table 1-4 lists the approved prefix names and symbols. Prefix symbols are written in roman type.

The kilogram is the only unit whose name and symbol, for historical reasons, both include a prefix. Names and symbols for decimal multiples and submultiples of the unit of mass are formed by attaching prefix names to the unit name "gram," and prefix symbols to the unit symbol "g", for example, "mg" is the symbol, rather than "μkg".

● 1.3 Traceability and measurement standards

To make a reliable measurement using a measuring instrument, it is essential to calibrate the instrument by a measurement standard that features higher accuracy or lower measurement uncertainty. It is then necessary to calibrate the measurement standard used by another standard with much higher accuracy. This procedure is repeated until the calibration is carried out at a national or an international standard and the measurements provided by the measuring instrument can be recognized as traceable to that national or international standard. Metrological traceability is

定義されている．

国家標準は，国家または経済圏で使用するために国家当局が承認した測定標準であり，その量がかかわるその他の測定標準にその量の値を供給する基礎となる．日本の場合，国家標準となる特定標準器，特定標準物質等は経済産業大臣が指定する．国際標準は国際協定の署名者によって承認され，世界中で用いられることを意図した測定標準のことである．質量を計測する秤が実用標準器の分銅，常用参照標準器の分銅，二次標準器の標準分銅，日本国キログラム原器の標準分銅を経て，キログラムの定義に辿り着くかを示す質量に関するトレーサビリティの連鎖を Fig. 1-1 に示す．世界中どこでも 1 kg が同じ質量として認められるのはキログラムの定義を頂点とするトレ

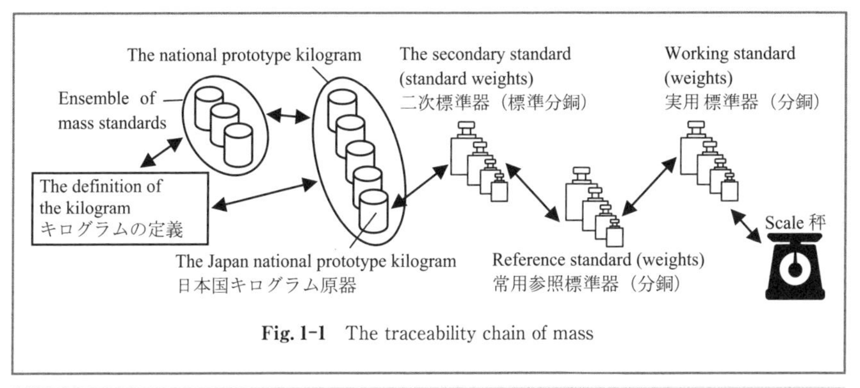

Fig. 1-1　The traceability chain of mass

defined as the property of a measurement result whereby the result can be related to a national or an international standard through a documented, unbroken chain of calibrations, each contributing to measurement certainty.

A national measurement standard is a standard recognized by a national authority to serve within a state or economy as the basis for assigning quantity values to other measurement standards for the kind of quantity concerned. In Japan, such standards are recognized by the Minister of Economy, Trade and Industry. An international measurement standard is a measurement standard recognized by signatories to an international agreement, and is intended to serve worldwide. Fig. 1-1 shows the traceability chain of mass, which connects a scale to a weight of the practical standard of mass, a standard weight of the specified secondary standard of mass, and a primary standard weight of the specified primary standard of mass, the definition of the kilogram. Because of the traceability chain of mass, 1 kg can be recognized as the same amount of mass anywhere around the world.

A measurement standard is the realization of the definition of a given quantity,

ーサビリティがあるからである.

また,計測標準は物理量をその定義に基づいて具現化するものである.具現化の手順として,まずは定義に基づく測定単位の物理的具現化が挙げられる.次の手順は「再現」と呼ばれ,物理現象に基づいた再現性の高い測定標準を構築することで,例えば,長さ(メートル)の測定標準を確立するための周波数安定化レーザ,電圧(ボルト)のためのジョセフソン効果,または電気抵抗(オーム)のための量子ホール効果の利用などの場合にみられる.第三の手順は実量器を測定標準として採用することである.ただし,SI 単位系および計測標準は科学技術の進歩に伴って変化するものである.本章では第 26 回国際度量衡総会(CGPM)で承認された最新のもので記述している.

with a stated quantity value and associated measurement uncertainty, used as a reference by a measuring system, a material measure, or in reference material. The term "realization" is used here in its most general sense. It denotes three procedures of realization. The first consists of the physical realization of the measurement unit from its definition and is realization in a strict sense. The second, termed "reproduction", consists not of realizing the measurement unit from its definition but in setting up a highly reproducible measurement standard based on a physical phenomenon, as happens for example, in the use of frequency-stabilized lasers to establish a measurement standard for the meter, of the Josephson effect for the volt, or of the quantum Hall effect for the ohm. The third procedure consists in adopting a material measure as a measurement standard. It should be noted that the SI systems and measurement standards can change with the progress of science and technology. In this chapter, the new definitions of the SI base units, approved at the 26 th General Conference on Weights and Measures (CGPM), are presented.

【演習問題】

1-1）次の量の次元を式(1-2)のように示せ：加速度，トルク，角度振動数

1-2）13.46 mm は何 pm であるか.

1-3）2019 年に改正される前，SI 基本単位の中で唯一人工物により定義されていたのは何か.

1-4）単位記号はどの書体で表すか.

1-5）1960 年までメートルを定義していた基準の名前を示せ．現在の定義に変わった理由を吟味せよ.

【Problems】

1-1) Show the dimensions of acceleration, torque and angular frequency as in Eq. (1-2).

1-2) How many pm is 13.46 mm?

1-3) There was only one SI base unit defined by an artifact before SI was revised in 2019. What was it?

1-4) What is the typeface for a unit?

1-5) Show the name of the standard for defining the unit of meter before 1960. Explain the reason it has been changed to the current definition.

第2章　計測システムの構成と特性

計測には必ず計測機や計測システムが必要となる．計測システムを用いて計測を行う際，計測システムの構成を理解した上で，その特性を把握しておくことが重要である．本章では，計測システムの一般的な要素を示し，その特性を評価するための項目について述べる．

● 2.1　計測システムの構成

ユーザが計測システムあるいは計測機を用いて計測を行う場合は，Fig. 2-1 に示すように，まず入力となる計測対象の物理量（計測量）をセンサで電気信号に変換する．

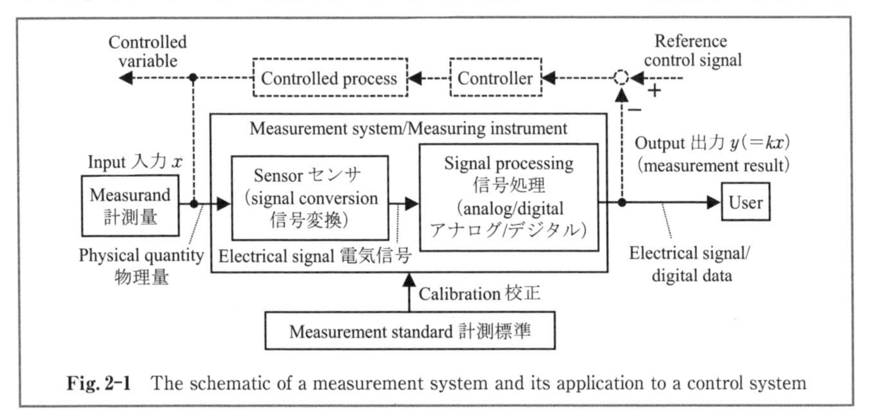

Fig. 2-1　The schematic of a measurement system and its application to a control system

Chapter 2　Configuration and Characteristics of Measurement Systems

A measuring instrument or measurement system is essential for measurement. It is important to understand the configuration and the characteristics of a measurement system when using it for measurement. In this chapter, the typical elements of a measurement system are presented and the evaluation parameters are described.

● 2.1　Configuration of a measurement system

As Fig. 2-1 shows, the input of a measurement system is the measurand, which is the physical quantity to be measured. The measurand is converted into an electrical

センサは測定システムの中核的な要素であり，何らかの科学的原理や物理法則に基づいて，計測対象の物理量を人間や機械が取り扱いやすい電気信号に変換する信号変換器の役割を果たしている．また，センサは計測対象と計測システムをつなぐインターフェースでもある．センサは人間の五官を代替するものとして発展してきたといえる．Table 2-1 にそれらのセンサの一部を示す．一方，磁気など，人間では全く感じることのできない量を測定できたり，非可視光や超音波など，人間の五官で感じられる範囲をはるかに超える領域をカバーしたりするなど，人間の五官の能力を超えるセンサも多く開発されている．本書では，機械系でよく用いられる物理量センサを中心に取り上げる．

その電気信号を計測システムの出力としてそのまま出力するケースもあるが，多くの場合は後続電気回路で増幅やフィルタリングなどの処理を行い，計測量に比例する

Table 2-1　The sensors to replace human sensory organs

Sensory organs 五官	Eyes 目	Ears 耳	Tongue 舌	Nose 鼻	Skin 皮膚
Senses 五感	Sight 視覚	Hearing 聴覚	Taste 味覚	Smell 嗅覚	Touch 触覚
Quantities 量	Light 光 (Physical)	Vibration 振動 (Physical)	Chemical substances 化学物質 (Chemical)	Chemical substances 化学物質 (Chemical)	Force 力 Pressure 圧力 Displacement 変位 Temperature 温度 (Physical)
Sensors センサ	Optical sensors 光センサ	Pressure sensors 圧力センサ	Taste sensors 味覚センサ	Gas sensors ガスセンサ	Force 力センサ Pressure 圧力センサ Displacement 変位センサ Temperature 温度センサ

quantity by a sensor based on scientific principles and/or physical laws. The sensor is a key component of the measurement system. A sensor also acts as the interface connecting the measurand and the system. Sensors have been developed to replace the sensory organs of human beings. Table 2-1 shows some such sensors. Many sensors even have capabilities and offer performances that are superior to human abilities. For example, magnetic sensors can sense magnetism that cannot be sensed by any human sensory organ. Optical sensors have a wide spectrum range that includes violet lights and infrared lights, which are not visible to human eyes. In this book, sensors that are often used in mechanical engineering will be presented.

The electrical signal from a sensor is often modified by succeeding circuits like amplifiers or filters before it is output by the measurement system as the

電気信号を測定結果として出力したり，あるいはセンサからの電気信号を数値データに変換してマイコンなどで必要な数値演算処理をしてから測定量に比例する同じ量の数値データを測定結果として出力することとなっている．ユーザのほうでこの測定結果を製品の品質管理などにそのまま利用したり，あるいは，コントローラを介して，機械や装置の制御に用いることができる．

計測システムを利用して計測を行うときは，求められる計測結果と計測量の性質に合わせて適切な特性を持つ計測システムを選ぶ，あるいは開発する必要がある．以下では，計測システムの特性を静特性と動特性に分けて説明した後，計測システムを用いて得られた計測結果の基本特性について述べる．

● 2.2　計測システムの静特性

測定対象の計測量と測定結果はそれぞれ計測システムの入力と出力となり，入力計測量 x と出力測定結果 y とは Fig. 2-2 の点線で示す $y=kx$ のような線形関係が理想的である．ここでは x と y の関係を計測システムの入力－出力関係と呼ぶ．入力の時間的変化を考えない場合の計測システムの入力－出力特性を静的特性と呼び，主な

measurement result. In some cases, the electrical signal is converted into digital data to obtain the measurement result of the system, through the use of numerical data processing. The measurement result can be utilized for the quality control of products. It can also be employed for feedback control of a machine or a device through a controller.

When using a measurement system for a specific measurement, it is necessary to select or develop a measurement system with proper characteristics that can satisfy the requirements of that measurement. The characteristics of a measurement system can be classified into static characteristics and dynamic characteristics. The characteristics of the measurement result are also important. These characteristics are described in the following sections.

● 2.2　Static characteristics of a measurement system

The characteristics of the measurement system are referred to as static characteristics when the input measurand x and the output measurement result y do not change with time. Ideally, y is proportional to x with a sensitivity coefficient k, as shown by the dashed line in Fig. 2-2. The main features of static characteristics are as

項目は以下のとおりであり

■測定範囲（Δx）：測定システムが計測できる最大の入力範囲（計測量の範囲）であり，計測できる入力の最大値 x_{max} と最小値 x_{min} との差で表す．

■出力範囲（Δy）：測定システムの出力が変化する最大範囲であり，出力の最大値 y_{max} と最小値 y_{min} との差で表す．

■感度係数（k）：入力計測量の変化に対する測定システム出力の変化の割合であり，出力変化量をそれに対応する入力の変化量で割ったものとして求める．

■信号対雑音比（SNR）：計測システム内部の電子回路で生じる電気ノイズなどによって，計測システムの出力には雑音成分が含まれる．計測システムの出力範囲（Δy）と雑音の大きさ（Noi）との比を信号対雑音比（SNR）と呼ぶ．一般的には，比の 20 log をとり，dB を単位として表す．

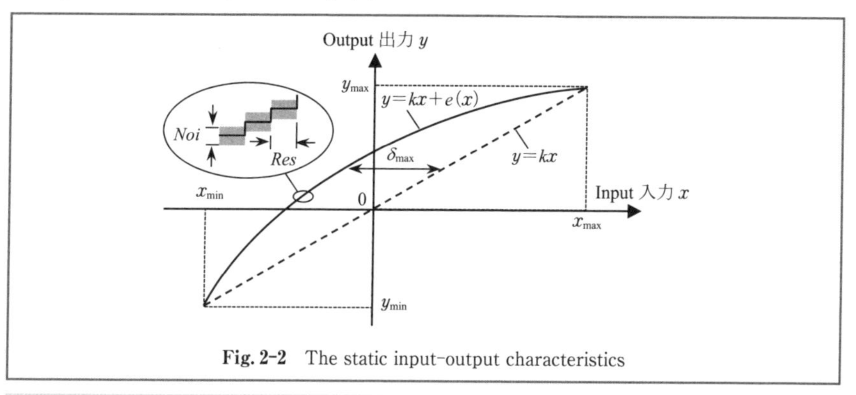

Fig. 2-2　The static input-output characteristics

follows:

■ Measurement range (Δx): This is the maximum measurable range of the measurand, expressed by the difference between the maximum input x_{max} and the minimum input x_{min}.

■ Output range (Δy): This is the maximum range of the measurement result, expressed by the difference between the maximum output y_{max} and the minimum output y_{min}.

■ Sensitivity coefficient (k): This is the ratio of the output change to the input change. The mean sensitivity coefficient can be evaluated by dividing Δy by Δx.

■ Signal to noise ratio (SNR): Noise components are included in the output. SNR is the ratio of Δy to the amplitude (Noi) of the noise. The SNR unit is often changed to dB by applying 20 log to the ratio.

■ Resolution (Res): This is the minimum measurable change of the input measurand,

■分解能（*Res*）：測定システムで測定可能な最小入力変化である．分解能はセンサの測定原理や，処理回路のノイズレベルやデジタル演算処理のビット数などによって決まる．分解能を測定範囲に占める割合（a%）で表す場合もある．

■ダイナミックレンジ(D_R)：測定範囲Δxと分解能*Res*の比であり，一般的には比の20 logをとり，dBを単位として表す．とくに電圧などアナログ信号を出力とする測定システムでは，ダイナミックレンジがほぼ一定あり，測定範囲と分解能はトレードオフの関係にある．なお，ダイナミックレンジは信号対雑音比と対応している場合も多い．

■線形性誤差($\delta_{\max}$)：センサや処理回路などの特性によって，実際の計測システムの入力と出力はFig. 2-2の点線で示す理想的な線形関係からずれてしまうことが一般的である．この場合，出力には非線形的な成分$e(x)$が含まれることとなる．線形性誤差は図に示す最大ずれ量$\delta_{\max}$で表し，それが測定範囲Δxに占める割合（b%）で示す場合も多い．

Fig. 2-2では二次の非線形成分が含まれるように示されているが，実際の計測システムの場合は，より高次の非線形性成分が存在する場合が多い．なお，Fig. 2-2の測

which is determined by the sensor principle, the noise level of the signal conditioning circuits, the least significant bit (LSB) of the numerical data processing, etc. The resolution is sometimes expressed as a percentage (a%) of the measurement range Δx.

■ Dynamic range (D_R): This is the ratio of the measurement range Δx to the resolution *Res*. The D_R unit is often changed to dB by applying 20 log to the ratio. For a measurement system with an analog output such as voltage, D_R scarcely changes with a change in Δx, so Δx and *Res* are in a trade-off relationship, which means that a larger Δx implies a lower *Res*. D_R often corresponds to *SNR*.

■ Linearity error ($\delta_{\max}$): In actual cases, the relationship between the input measurand and the output measurement result will deviate from the ideal relationship shown by the dashed line in Fig. 2-2, due to the sensor principle and the data conditioning circuits. In this case, a non-linear component $e(x)$ will be included in the output. The maximum $\delta_{\max}$ of the deviation is referred to as the linearity error, which is often expressed as a percentage (b%) of the measurement range Δx.

For clarity, in Fig. 2-2 a second-order non-linear error component is shown in the input-output curve. It should be noted that in actual cases, higher-order components are included. The non-linear input-output curve shown in Fig. 2-2 can be identified

定システムの入力－出力特性曲線はより精度の高い計測基準を用いた校正で得ることができ，校正結果から感度係数や線形誤差などの特性を評価することができる．また，校正結果に基づいて，計測システムの特性を補正することも可能である．

計測システムの特性は Fig. 2-2 の符号を用いて表現すると Table 2-2 のようになる．また，Table 2-2 には，市販の静電容量型変位センサの例も合わせて示している．

Table 2-2 Static characteristics of a measurement system

Parameters 項目	Expressions 定義式	Example of a commercial displacement sensor 変位センサ製品例
Measurement range 測定範囲	$\Delta x = x_{\max} - x_{\min}$	$x_{\max} = 50\ \mu\text{m}$, $x_{\min} = -50\ \mu\text{m}$ $\Delta x = 50\ \mu\text{m} - (-50\ \mu\text{m}) = 100\ \mu\text{m}$
Output range 出力範囲	$\Delta y = y_{\max} - y_{\min}$	$y_{\max} = 10\ \text{V}$, $y_{\min} = -10\ \text{V}$ $\Delta y = 10\ \text{V} - (-10\ \text{V}) = 20\ \text{V}$
Sensitivity coefficient 感度係数	$k = \dfrac{\Delta y}{\Delta x}$	$k = \dfrac{20}{100} = 0.2\ \text{V}/\mu\text{m}$
Signal noise ratio SN 比 (dB)	$SNR = 20 \log \dfrac{\Delta y}{Noi}$	$SNR = 20 \log \dfrac{20}{0.002} = 80$
Resolution 分解能	$Res = \Delta x / 10^{\frac{D_R}{20}}$ or $Res = a\% \times \Delta x$	$Res = 0.01\% \times 100 = 0.01\ \mu\text{m}$
Dynamic range ダイナミックレンジ(dB)	$D_R = 20 \log \dfrac{\Delta x}{Res}$	$D_R = 20 \log \dfrac{100}{0.01} = 80$
Linearity error 線形誤差	$\delta_{\max} = b\% \times \Delta x$	$\delta_{\max} = 0.1\% \times 100 = 0.1\ \mu\text{m}$

by using a calibration process with a reference system of higher accuracy. The obtained calibration result can also be employed to compensate for the non-linear component.

Table 2-2 summarizes the static characteristics of the measurement system, using the symbols from Fig. 2-2. As an example, the characteristics of a commercially available capacitance displacement sensor are also shown in the table.

● 2.3　計測システムの動特性

時間 t とともに変化する計測量 $x(t)$ を計測する場合は，計測システムの動特性を考える必要がある．実際の計測システムではセンサおよび信号処理回路の周波数特性によって，$x(t)$ と計測システムの出力 $y(t)$ の間には単純な比例関係が成立しない．その場合，Fig. 2-3 と 2-4 に示すように，出力 $y(t)$ と入力 $x(t)$ の振幅の比（ゲイン）および位相差は入力の周波数 f あるいは角周波数 ω（$=2\pi f$）の関数となる．Fig. 2-4 のボード線図で表すゲインおよび位相差と周波数の関係を計測システムの周波数特性という．

計測システムの周波数特性の表現には，伝達関数が有効的である．入力をインパルス関数としたときの計測システムの出力を $g(t)$ とする．$G(j\omega)$ は $g(t)$ をフーリエ変

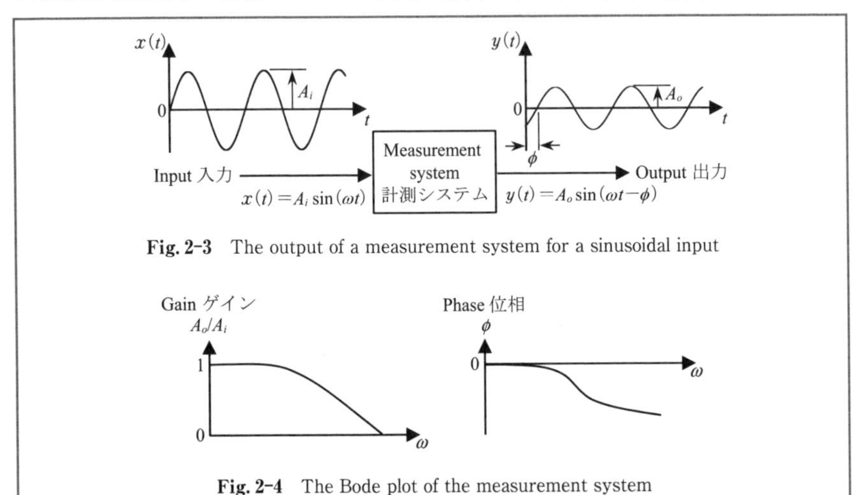

Fig. 2-3　The output of a measurement system for a sinusoidal input

Fig. 2-4　The Bode plot of the measurement system

● 2.3　Dynamic characteristics of a measurement system

The dynamic characteristics of a measurement system are important when the system is used to measure a time-dependent measurand $x(t)$, where t is time. In an actual measurement system, $y(t)$ is not simply proportional to $x(t)$ due to the time-dependent issues in the sensor principle and the signal conditioning circuits. As Figs. 2-3 and 2-4 show, both the gain, which is the ratio of the amplitude of $y(t)$ to that of $x(t)$, and the phase difference between $y(t)$ and $x(t)$ are functions of the frequency f or the angular frequency ω ($=2\pi f$) of the input. The relationships of the gain and the phase with respect to the frequency, which are shown in the Bode plot of Fig. 2-4, are referred to as the frequency characteristics of a measurement system.

It is effective to identify the frequency characteristics of a measurement system

換して得たものであり，周波数域における計測システムの伝達関数となり，周波数伝達関数という．$G(j\omega)$ は計測システムの周波数特性そのものを表している．Fig. 2-5のように，時間域では，計測システムの出力は入力とインパルス応答の畳み込み積分という複雑な関係式となっているが，周波数域では，出力は入力の変換 $X(j\omega)$ と $G(j\omega)$ の積という単純な関係式で表せるので，$G(j\omega)$ を用いて計測システムの動的特性を表現するのは適切であるといえる．式(2-1)のように，$|G(j\omega)|$ と $\phi(\omega)$ はそれぞれ計測システムのゲイン特性と位相特性を表している．ω の範囲を広くとると $|G(j\omega)|$ が大幅に変化するので，一般的には $|G(j\omega)|$ の代わりに $20\log|G(j\omega)|$ をゲインとして用いている．

一方，時間域において，計測のシステムの入出力関係は式(2-2)のような n 次常微分方程式で表すことができ，その周波数伝達関数 $G(j\omega)$ は式(2-3)のように表せる．$n=0$ の0次システムにおいては，出力と入力は理想的な比例関係にあり，その時 a_0 の逆数は Fig. 2-2 の感度係数 k と等しくなる．また，$n=1$ の一次システムにおいて

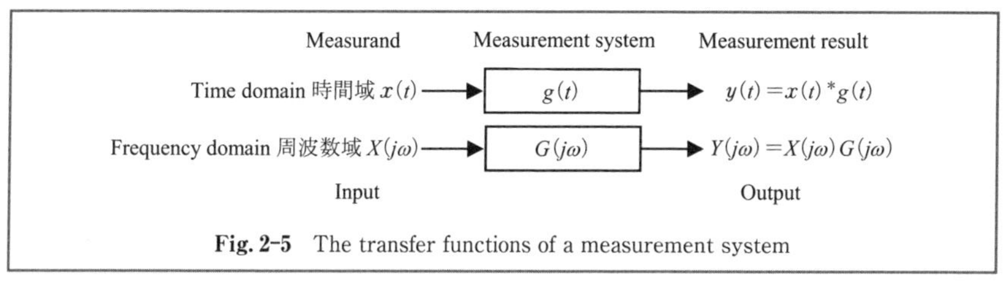

Fig. 2-5　The transfer functions of a measurement system

by using the transfer function. Assume the output of the system is $g(t)$ for an impulse input in Fig. 2-5. $G(j\omega)$, which is the Fourier transform of $g(t)$, demonstrates the transfer function of the system in the frequency domain and is called the frequency transfer function. As shown in the figure, the output of the system is a complicated convolution function of the input and the impulse response in the time domain. On the other hand, the output becomes a simple product of the Fourier transform $X(j\omega)$ of the input and the frequency transfer function $G(j\omega)$ in the frequency domain. $G(j\omega)$ is expressed by Eq. (2-1), where $|G(j\omega)|$ and $\phi(\omega)$ are the gain characteristics and the phase characteristics respectively. Since $|G(j\omega)|$ can change greatly over a wide range of ω, $20\log|G(j\omega)|$ is typically employed instead of $|G(j\omega)|$ to express gain characteristics.

It is also possible to express the input and output relationship by the n th ordinary differential equation in the time domain in Eq. (2-2), with its frequency transfer function shown in Eq. (2-3). For a zeroth order system of $n=0$, the input and the output have an ideal proportional relationship and the reciprocal of a_0 corresponds to

は，$a_0=1$ の場合，$\omega=1/a_1$ のときに $|G(j\omega)|$ が $1/\sqrt{2}\approx0.707$ となり，$20\log|G(j\omega)|$（ゲイン）が約 -3 dB となる．この時の周波数を遮断周波数（カットオフ周波数）といい，ゼロから遮断周波数までの周波数範囲を計測システムの周波数帯域と呼ぶ．一次システムでは，ω_cにおける位相 ϕ は $-45°$ となる．なお，単位ステップ入力については，$X(j\omega)$ は $1/j\omega$ であり，$G(j\omega)X(j\omega)$ から求めた $Y(j\omega)$ の逆フーリエ変換で計測システムのステップ時間応答を求めることができる．Fig. 2-6 に示すのは一次システムの単位ステップ応答である．この図から，システムの動的特性を表すパラメータとしての時定数や遅れ時間，定常偏差などを読み取ることができる．

$$G(j\omega)=a(\omega)+jb(\omega)=|G(j\omega)|e^{j\phi(\omega)}=|G(j\omega)|\angle\phi(\omega)$$

$$\text{where }|G(j\omega)|=\sqrt{a^2(\omega)+b^2(\omega)},\qquad \phi(j\omega)=\angle G(j\omega)=\tan^{-1}\frac{b(\omega)}{a(\omega)} \tag{2-1}$$

$$a_n\frac{d^ny(t)}{dt}+a_{n-1}\frac{d^{n-1}y(t)}{dt}+\cdots+a_1\frac{dy(t)}{dt}+a_0y(t)=x(t) \tag{2-2}$$

$$G(j\omega)=\frac{Y(j\omega)}{X(j\omega)}=\frac{1}{a_n(j\omega)^n+a_{n-1}(j\omega)^{n-1}+\cdots+a_1(j\omega)^1+a_0} \tag{2-3}$$

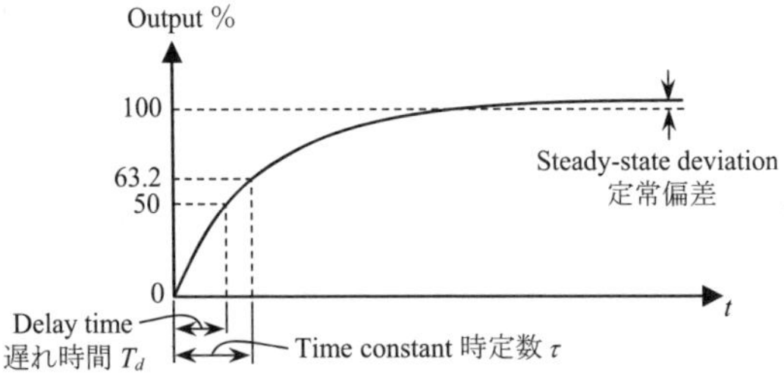

Output %
100
63.2
50
0
t
Steady-state deviation
定常偏差
Delay time
遅れ時間 T_d
Time constant 時定数 τ

Fig. 2-6 The unit step response of a 1st-order measurement system

the sensitivity coefficient k in Fig. 2-2. For a first-order system of $n=1$, $|G(j\omega)|$ is $1/\sqrt{2}\approx0.707$ and $20\log|G(j\omega)|$ is approximately -3 dB; and the phase ϕ is $-45°$ when $\omega=1/a_1$ and $a_0=1$. This frequency is referred to as the cutoff frequency. The frequency range from zero to the cutoff frequency is called the frequency bandwidth of the system. For a step input, $X(j\omega)$ is $1/j\omega$ and $Y(j\omega)$ can be evaluated from $G(j\omega)X(j\omega)$. Therefore, the response of the system to the step input in the time domain can be obtained from the inverse Fourier transform of $Y(j\omega)$. Fig. 2-6 shows such a unit step response of a first-order system, from which several important dynamic characteristics of the system, such as the time constant, the delay time, and the steady-state deviation, can be obtained.

● 2.4　計測結果の基本特性

計測システムを用いて計測を行う場合，計測システムの特性のみならず，計測者や計測環境などにも影響され，計測値がばらつく現象が必ず生じる．そのため，同じ測定を複数回行い，平均値をとることによってばらつきを低減させ，計測結果の信頼性を向上させる必要がある．Fig. 2-7 に同じ計測を N 回反復した場合の計測結果の特性を示す．計測値 $y(i)$ はその平均値 y_M を中心にばらついていることが分かる．図には反復計測値のばらつきが正規分布をしている場合の度数分布曲線も示している．反

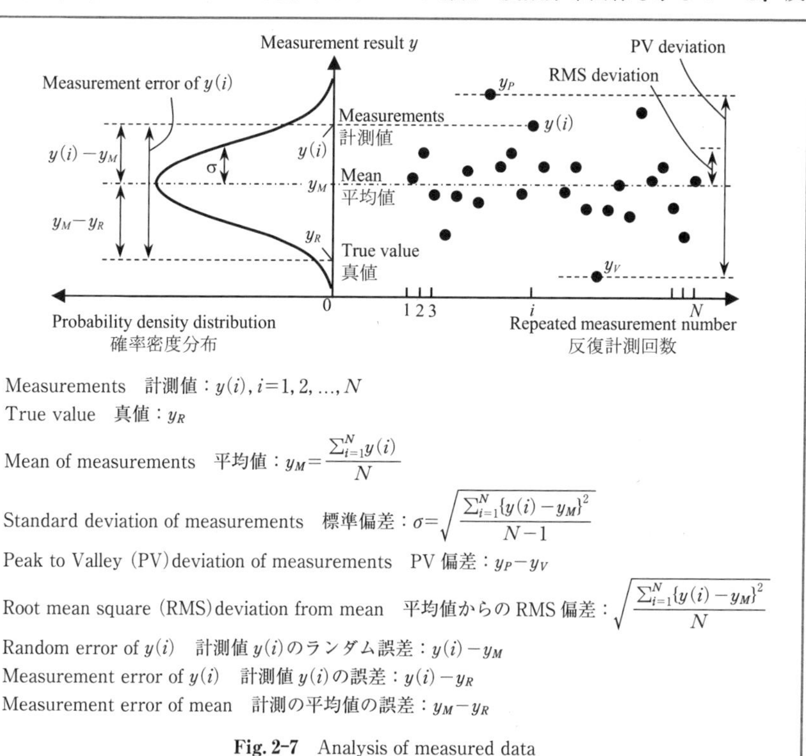

Fig. 2-7　Analysis of measured data

● 2.4　Characteristics of the measurement result

The measured data will disperse due not only to the characteristics of the measurement system but also to the operator and the environment. For this reason, it is necessary to repeat the measurement and to take the average of the measured data as the measurement result, so that the dispersions of data can be reduced for a higher reliability of measurement. As shown in Fig. 2-7, the measured data $y(i)$ of N replicated measurements disperse around the mean value y_M. The probability density

復計測値のばらつきの程度を標準偏差σで評価することができる．標準偏差σとほぼ同じ値となるが，計測機出力のノイズレベルなどの評価には二乗平均平方根(RMS) 偏差を用いる場合も多い．

計測値$y(i)$の計測誤差は計測値と真値y_Rとの差で定義されており，系統誤差成分と偶然誤差成分の和になっている．系統誤差$y(i)$の平均値とy_Rの差，また偶然誤差は$y(i)$とy_Rの差でそれぞれ求められる．系統誤差の大きさはほぼ一定であるか，その変化が予測できるのに対して，偶然誤差は反復計測の度にその大きさが予測できない変化をするという特徴がある．反復計測の平均値を計測結果として用いたほうが，単独の計測値を用いるよりも，計測誤差が小さくなるといえる．

また，Fig. 2-8 に示すように，偶然誤差と系統誤差は計測精度の成分である精密さと正確さにそれぞれ対応している．計測の精密さの程度を表すパラメータとして繰り返し性と再現性がある．繰り返し性は計測手順，測定者，計測機，測定場所などが同

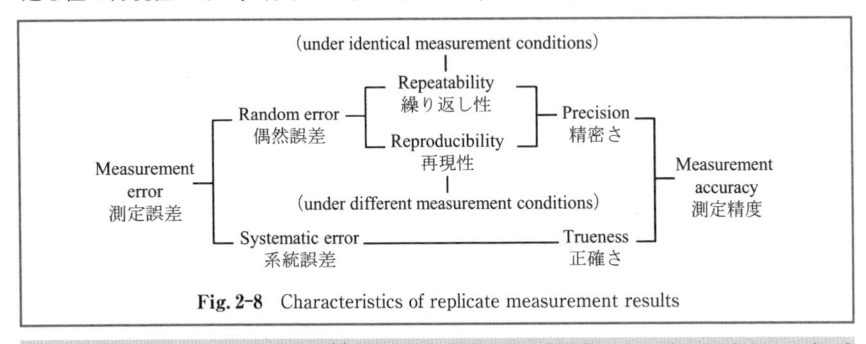

Fig. 2-8　Characteristics of replicate measurement results

for a normal distribution of $y(i)$ is also shown in the figure, where the standard deviation σ is employed for evaluating the dispersion of $y(i)$. On the other hand, the root mean square (RMS) deviation, which is almost equal to the standard deviation, is often used for evaluation of the noise component in measurement data.

The measurement error, which is composed of the systematic error component and the random error component, is the difference between the measured value and the true value y_R. The systematic error, which is almost constant or has predictable changes, is the difference between the mean of $y(i)$ and y_R. The random error is the difference between $y(i)$ and y_R. It varies whenever the measurement is repeated and the change is unpredictable. This error can be reduced by taking the mean of the replicated measured data as the measurement result.

As Fig. 2-8 shows, the systematic error and the random error correspond to trueness and precision from the point of view of accuracy. The degree of precision can be expressed by repeatability and reproducibility. Repeatability is precision under the

じ測定条件下で得られた反復計測値のばらつきの度合いを示していることに対して，再現性は測定条件を全部あるいは一部変更した条件下で得られた反復計測値のばらつきの度合いを示している．

計測精度および計測誤差は計測値と真値との差で定義されているが，本当は真値が分からないので，この定義は定性的なものになっている．真値を必要としない定量的な計測値信頼性評価パラメータとして，次章で述べる計測の不確かさがある．

【演習問題】

2-1）計測システムの精度と分解能の値はどちらが大きいか．

2-2）式(2-2)で示す一次システムの単位ステップ応答を式で表し，それに基づいて，一次システムの時定数 τ と遅れ時間 T_d を求めよ．

2-3）ある測定量に対して同じ計測を 10 回反復して得た測定値は 5.12, 5.33, 5.19, 5.42, 5.25, 5.10, 5.28, 5.40, 5.11, 5.37 となっている．その計測値の標準偏差 σ を求めよ．

2-4）測定者 A と B で同じ計測量に対して行った反復測定の分散は再現性と繰り返

same set of measurement conditions such as procedure, operator, measuring system and location. Reproducibility is precision under different conditions.

The measurement accuracy and the measurement error are defined with respect to the unknown true value and are thus qualitative. The measurement uncertainty, which is a quantitative definition that does not use the true value, is presented in the next chapter.

【Problems】

2-1) Which one has a larger value, the resolution or accuracy of a measurement system?

2-2) Express the equation for the unit step response of a first-order system shown in Eq. (2-2) and evaluate the time constant τ and the delay time T_d of the system based on the equation.

2-3) The replicated measurements on a measurand are 5.12, 5.33, 5.19, 5.42, 5.25, 5.10, 5.28, 5.40, 5.11, 5.37. Evaluate the standard deviation σ of the measurements.

2-4) Does the dispersion between the replicated measurements made on the same

し性のどちらになるか.

2-5) 正規分布を持ったノイズ成分の最大幅（PV 値）は二乗平均平方根（RMS 値）の約何倍であるか.

measurand by Operators A and B, respectively, correspond to reproducibility or repeatability?

2-5) What is the approximate ratio of the peak-to-valley (PV) value to the RMS value of the noise component with a normal distribution in the measured data?

第3章　計測の不確かさ

何かしらの物理量をはかる場合，それにより得られる結果には，真の値とのずれ（誤差）が必ず含まれる．第2章で述べたとおり，計測誤差は，真の値からの一定の偏りである系統誤差成分と，測定時の偶発的要因により発生する偶然誤差成分に分類される．この系統誤差・偶然誤差の考え方は，実際の現場での要因分析に有用であることもあり，これまで広く用いられてきた．これに対し，近年では「計測の不確かさ」という概念の導入が進められている（Fig. 3-1）．以下では，この計測の不確かさの概念について説明するとともに，マイクロメータを用いた長さ測定を例として挙げ，計測の不確かさ分析の手順について述べる．

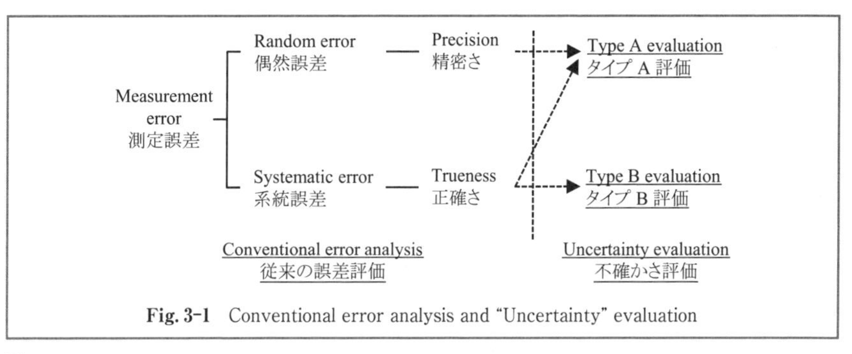

Fig. 3-1　Conventional error analysis and "Uncertainty" evaluation

Chapter 3　Measurement Uncertainty

A measurement result always contains a deviation (error) from the value of a measurand. As described in Chapter 2, measurement error consists of systematic and random components, which are constant and random deviations from the measurand's value respectively. This categorization has been employed thus far due to its usefulness for error analysis. However, its definition is based on the value of a measurand that cannot be measured. Furthermore, it is difficult to establish a method to evaluate variations in repeated observations in most cases. Therefore, the concept of "measurement uncertainty" has recently been introduced (see Fig. 3-1). The concept of measurement uncertainty is described in what follows. A procedure for

● 3.1　不確かさの概念

計測において最終的に求めたい真値は，実際には求めることができず，最終的に得られる「確からしい値」には必ず誤差が含まれる．真値は，得られる「確からしい値」を含むある一定の範囲に存在するが，その一定の範囲を得るためには，偏り（系統誤差）とばらつき（偶然誤差）を合成して評価する必要がある．しかし，その合成手法は一義に定まっておらず，また，事象によっては系統誤差とも偶然誤差ともとれる場合がありその判断が難しい場合もあるなど，その取扱いについて普遍的な手法が確立できていないという問題があった．さらに，誤差の概念は，実際には求めることができない真値ありきで定義されており，その定義そのものに曖昧さが残ることとなる．

このような定義の曖昧さを排除するとともに，測定結果の妥当性を合理的かつ定量的に評価するための新たな概念として，「計測の不確かさ」の概念が導入された．1993年には，計測にかかわる主要な7国際機関による共同編集で"*Guide to the Expression of Uncertainty in Measurement*"（GUM）（測定における不確かさの表現ガイド）が発行された．GUMにおいては，測定の不確かさ要因を以下の2つの形式に分

uncertainty analysis is also explained in reference to measuring length with a micrometer.

● 3.1　Concept of measurement uncertainty

The value of a measurand cannot be known in practice, and the result of any measurement, which is only an approximation of the value of the measurand, always contains an error. The value of a measurand therefore always exists in a certain range, which must be acquired by combining both the random and systematic components of the error. However, a foolproof method of combining these error components has not yet been established, since it is sometimes difficult to distinguish whether a component is random or systematic. Furthermore, the definitions of both random and systematic components are vague, since both are based on the variable of a measurand, which is an unknowable quantity.

To eliminate this ambiguity and to evaluate the validity of measurement results reasonably and quantitatively, the concept of "measurement uncertainty" has been introduced. In 1993, the "*Guide to the Expression of Uncertainty in Measurement*" (GUM) was published as a collaborative effort by the major seven international organizations involved in measurement. In the GUM, uncertainty components are

類して取り扱う．

・Type A：一連の測定値の統計的解析による不確かさの評価の方法

・Type B：一連の測定値の統計的解析以外の手段による不確かさの評価の方法

これら Type A および B 評価によって，各要因による測定の不確かさを標準偏差の形で表現し，不確かさの伝播則を適用して合成することで，総合的な測定不確かさを定量的に表現する．統計的解析であるか否かによって機械的に Type A および B 評価に分類し，かつその合成の際にはタイプに依らず同等に扱うため，作業者の恣意的な意図が入り込む余地がなく，結果として合理的な測定結果の評価が実現する．

3.2　計測における不確かさの定量的評価方法

GUM による計測の不確かさ評価では，測定方法を数学モデルによって表現し，それに含まれる各変数が計測結果に及ぼす影響を算出する．その算出過程においては，標準偏差 σ での表現が基本となる．標準偏差 σ は元のデータと同じ次元を有し，ばらつきの指標として都合が良い．GUM では，不確かさの伝播則をベースとして，

grouped into two categories based on their methods of evaluation:

· Type A: An uncertainty value estimated from the statistics of a variable

· Type B: An uncertainty value estimated by other than statistical methods

The measurement uncertainty due to each uncertainty component is expressed in a standard deviation. By applying the concept of error propagation, the estimated standard uncertainties will be combined as a sum of these squares to express measurement uncertainty quantitatively. Since the uncertainty components will be categorized only by considering their methods of evaluation, and all the uncertainty components will be combined regardless of their uncertainty types, reasonable uncertainty evaluation can be realized while eliminating any intentions of human operators.

3.2　Measurement uncertainty analysis

In the GUM, measurement is expressed by using a mathematical model containing several input estimates that affect a measurement result. In the calculation procedure, each uncertainty component is expressed in the form of a standard deviation σ. σ has the same dimension as the measurement result, and is appropriate for expressing the degree of deviation. In the GUM, based on the concept of the error

個々の不確かさ成分を標準偏差で表し，これを二乗和で合成して表現する．

計測の数学モデル構築 まず，準備段階として，計測に用いる計測機についての情報を整理する．計測機の緒元表や試験成績書，校正証明書などをそろえる．その後，得られた情報をもとに，測定の数学モデルを検討する．計測手順，計測の定義を明確にするとともに，入力量 x_i(i は自然数) をリストアップし，すべての入力量を包含する関数 $f(x_1, x_2, \ldots, x_i, \ldots, x_n)$ として計測の数学モデルを表現する．

各入力量に対する標準不確かさの評価 数学モデルが構築できたら，各入力量 x_i に対する標準不確かさ $u(x_i)$ を算出する．

■ Type A 評価の場合：実験で得られたデータを用い，その標準偏差 σ から標準不確かさ $u(x_i)$ を算出する．なお，標準偏差が σ である測定を n 回繰り返し，その平均値を測定値として採用する場合の不確かさ $u(x_i)$ は式(3-1)より得られる．

$$u(x_i)=\frac{\sigma}{\sqrt{n}} \quad (3\text{-}1)$$

$$u(x_i)=\frac{a}{A} \quad (3\text{-}2)$$

$$c(x_i)=\frac{\partial f}{\partial x_i} \quad (3\text{-}3)$$

$$u_c=\sqrt{\sum_{i=1}^{n}\{c(x_i)u(x_i)\}^2} \quad (3\text{-}4)$$

$$U=k\cdot u_c \quad (3\text{-}5)$$

propagation, each uncertainty component in the standard deviation will be combined as a square root of the sum of the squares.

Modeling the measurement First, summarize all the information of the measuring instrument to be used. Gather related documents such as a specification sheet or a calibration certificate for the instrument. Then, establish a mathematical model; clarify the procedure and definition of measurement, while listing all the input values x_i (where i is a natural number) to express the measurement as a function $f(x_1, x_2, \ldots, x_i, \ldots, x_n)$.

Evaluation of the standard uncertainty associated with each input value Following the measurement modeling, estimate the standard uncertainty $u(x_i)$ of a measurand associated with each input value x_i.

■ For Type A uncertainty components: Calculate the standard uncertainty $u(x_i)$ by using the standard deviation σ of repeated independent observations. When the arithmetic mean of n repeated independent observations is employed as the result of measurement, the standard uncertainty $u(x_i)$ is calculated by Eq. (3-1).

■ Type B 評価の場合：測定に関する過去の経験，校正証明書，製造者の仕様書，常識から得られるもの，などを利用して不確かさを算出する．仕様書などで，不確かさが明記されている場合には，そのままその値を採用することができる．ただし，3σ で与えられている場合などは標準不確かさに換算して用いることに注意する．一方，不確かさの上限と下限しか推定できない場合には，その不確かさの分布を予想して，式(3-2)を用いて不確かさ $u(x_i)$ を算出する．式中，a は上限と下限の間の半値である．また A は不確かさ分布により決まる値であり，一様分布の場合は $\sqrt{3}$，正規分布の場合は 1，三角分布の場合は $\sqrt{6}$ となる（Fig. 3-2）．分布の扱いに迷った場合には，矩形分布として取り扱って差し支えないことが多い．

なお，$u(x_i)$ の次元は最終的に求めたい測定結果の合成不確かさと異なる場合がある．例えば，測長機による長さ測定において，環境温度 t のばらつきが測定結果に与

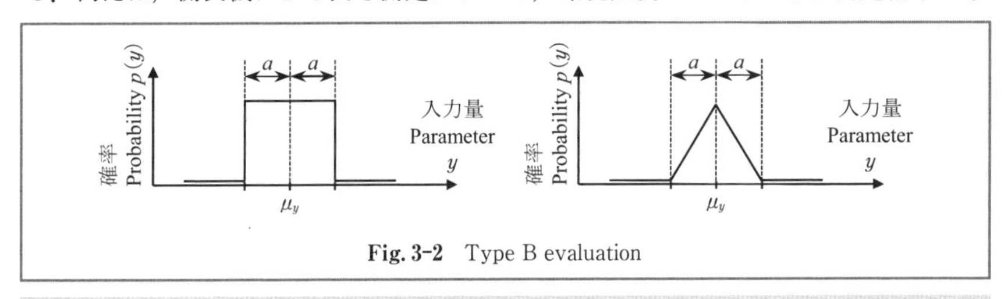

Fig. 3-2　Type B evaluation

■ For Type B uncertainty components: The standard uncertainty will be estimated based on information from past experiences, calibration certificates, specification sheets published by manufacturers, general knowledge, and so on. When the standard uncertainty is clearly described in such documents, that value can be directly employed as the estimated value. It must be noted that when the quoted uncertainty is stated to be a particular multiple of a standard deviation, the standard uncertainty $u(x_i)$ is simply the quoted value divided by the multiplier. When the input quantity x_i lies within the interval between $+a$ and $-a$, the standard uncertainty $u(x_i)$ can be estimated based on Eq. (3-2), while considering its probability distribution. In that equation, the parameter A is associated with the probability distribution; when the distribution can be treated as uniform, Gaussian, and triangular, the parameter A will be $\sqrt{3}$, 1, and $\sqrt{6}$, respectively (Fig. 3-2). When the probability distribution is unknown, the uniform distribution can be applied in most cases.

Sometimes the unit of a single standard uncertainty $u(x_i)$ differs from that of the combined standard uncertainty, as in the case of estimating the influence of the environmental temperature t on length measurement. In such cases, a sensitivity coefficient $c(x_i)$ will be included to describe how the output estimate varies with

える影響を考慮する場合などである．このような場合には，その次元の変換に感度係数 $c(x_i)$ を用いる．感度係数は，検討した数式モデルにおいて，各入力量 x_i が最終的な測定結果にどの程度の影響を与えるかの指標を示すものであり，検討した数式モデルを式(3-3)のように各入力量 x_i について偏微分することで容易に得られる．

合成不確かさの評価　こうして得られた各入力量 x_i に対する標準不確かさ $u(x_i)$ および感度係数 $c(x_i)$ をもとに，式(3-4)を用いて合成標準不確かさ u_c を計算する．なお，この式は不確かさの伝播則と呼ばれている．さらに，式(3-5)のように u_c に包含係数 k を乗ずることで拡張不確かさ U が得られる．

不確かさバジェット表の作成　得られた拡張不確かさの導出過程，および各入力量が計測の不確かさに与える影響を表にまとめる．

● 3.3　測定不確かさの評価例（マイクロメータの校正）

マイクロメータ（Fig. 3-3）は，物体の長さを簡便に精度よく測定する長さ測定機である．ここでは，校正証明書が付与された複数のブロックゲージを用いて校正した

changes in the input estimate x_i. The sensitivity coefficient is a parameter that describes the degree of the influence of the input estimate x_i on the combined standard uncertainty, and can be acquired as the partial derivative of a mathematical model with respect to x_i, as shown in Eq. (3-3).

Evaluation of the combined standard uncertainty　Based on the calculated standard uncertainties $u(x_i)$ and the sensitivity coefficients $c(x_i)$, calculate a combined standard uncertainty u_c by using Eq. (3-4), which is called a propagation of uncertainty. In addition, an extended uncertainty U can be acquired as the product of u_c and the coverage factor k, as shown in Eq. (3-5)

Build up the uncertainty budget table　Summarize the procedure of how to find each standard uncertainty and the combined uncertainty to clarify the influence of each input parameter on the measurement uncertainty.

マイクロメータの計測の不確かさを評価する．計測の前提条件は以下のとおりである．

・校正は室温 23.0℃（温度揺らぎ ±0.5℃）の部屋で行う．
・校正の対象は 0 mm～25 mm の測定レンジ
・マイクロメータの測定分解能は 0.001 mm
・校正に用いるブロックゲージの呼び寸法は 3 mm～25 mm
・いずれのブロックゲージにも 20℃環境下での校正証明書が添付されている．

計測の数学モデル構築 ここではマイクロメータの校正を，環境温度 20℃で保証されたブロックゲージの長さ(G_{20})を基準として実施する．いま，環境温度 t において

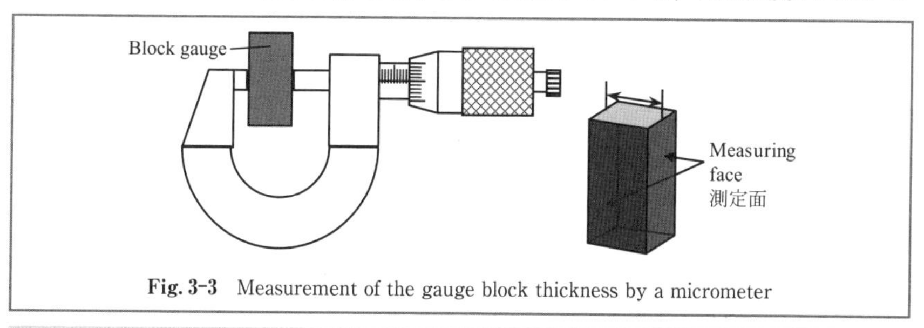

Fig. 3-3 Measurement of the gauge block thickness by a micrometer

3.3 An example of the evaluation of measurement uncertainty (calibration of a micrometer)

A micrometer (Fig. 3-3) is an instrument that can measure the length of a measurement target easily and precisely. What follows is an evaluation of the measurement uncertainty of a micrometer calibrated by using several block gauges with calibration certificates. We assume that the measurement will be carried out under the following conditions:

・Calibration is carried out in an air-conditioned room at a temperature of 23℃ (with a deviation of ±0.5℃).
・The measurement range runs from 0 mm to 25 mm.
・The measurement resolution of the micrometer is 0.001 mm.
・The nominal dimensions of the block gauges range from 3 mm to 25 mm.
・Each block gauge has a calibration certificate for temperatures under 20℃.

Modeling the measurement Calibration of the micrometer is carried out based on the lengths of the block gauges G_{20} certified at a temperature of 20℃. When the

マイクロメータを用いてブロックゲージを測定する場合，温度20℃における長さ G_{20} に比べて $(t-20)$℃の温度差で熱膨張した状態での長さが測られることになる．したがって，温度 t におけるマイクロメータの読み IR_t を，温度20℃の場合の読みに換算した後，その差をとって測定誤差 E_{20} を算出する必要がある．測定誤差の式は式(3-6)のように表せる．この式には以下4つのパラメータ x_i が含まれる．

・温度 t におけるマイクロメータの読み：IR_t

・ブロックゲージの熱膨張係数：α_{IR}

・環境温度 t とブロックゲージ校正時の温度(20℃)の差：Δt_{IR}

・環境温度20℃で保証されたブロックゲージの長さ：G_{20}

各入力量に対する標準不確かさの評価　次に，リストアップした変数 x_i それぞれについてその標準不確かさ $u(x_i)$ を評価する．

IR_t の不確かさは，以下の2要素を含む．

・マイクロメータの測定分解能：$u(Res)$

$$E_{20}=IR_t(1-\alpha_{IR}\Delta t_{IR})-G_{20} \tag{3-6}$$

micrometer is calibrated with the block gauges under temperature t, the lengths of the micrometer may change due to thermal expansion and will deviate with the temperature difference of $(t-20)$℃. Therefore, the reading of the micrometer IR_t at temperature t needs to be converted to the reading at 20℃ to calculate the measurement error E_{20}. E_{20} can be expressed as shown in Eq. (3-6), which contains the four input estimates x_i shown below:

・Reading the micrometer at temperature t: IR_t

・The thermal expansion coefficient of the block gauge: α_{IR}

・The temperature difference between the ambient temperature t and that at the calibration of the block gauge: Δt_{IR}

・Lengths of the block gauge calibrated at 20℃: G_{20}

Evaluation of the standard uncertainty associated with each input value

In the next step, evaluate the standard uncertainty $u(x_i)$ associated with each input parameter x_i listed in the procedure above.

The uncertainty of IR_t contains the two components listed below:

・The measurement resolution of the micrometer: $u(Res)$

・Measurement repeatability: $u(Rpt)$

・測定繰り返し精度：$u(Rpt)$

マイクロメータの測定分解能（目盛の最小値）による不確かさは，仕様表に記載の値から算出する（Type B 評価）．仕様書などに計測の不確かさが明記されている場合には，そのままその値を採用することができる．ただし，3σで与えられている場合などは標準不確かさに換算して用いることに注意する．一方，不確かさの上限$+a$と下限$-a$しか推定できない場合には，その不確かさの分布を予想して，式(3-2)をもとに不確かさ$u(x_i)$を算出する．いま，マイクロメータの測定分解能は仕様表より1 μm であることが分かっている．真値が目盛間のどの位置に存在するかは偶然により決まるため，その確率分布は矩形と考えられる．そのため，分解能の半値幅$a=0.5$ μm を適用して，$u(Res)=0.5\ \mu m/\sqrt{3}=0.289\ \mu m$ となる．

測定繰り返し精度$u(Rpt)$は，実際の実験により得られたデータを用いる（Type A 評価）．いま，長さ 25 mm のゲージを繰り返し測定した際の平均値は 25.00007 mm，標準偏差最大値は 0.577 μm であったとする．3 回の測定結果の平均値をマイクロメータの読みとして採用するものとして($n=3$)，式(3-1)より$u(Rpt)=0.577\ \mu m/\sqrt{3}=0.333\ \mu m$ が得られる．なお，このときの感度係数$c(IR_t)$は式(3-3)を用いて式(3-

A standard uncertainty due to the micrometer's measurement resolution can be estimated from the information in a specification sheet (Type B evaluation). When the measurement uncertainty is clearly indicated in a document, the indicated value can be directly applied for the evaluation purposes. Meanwhile, in the cases where the input quantity x_i lies within the interval from $+a$ to $-a$, the standard uncertainty $u(x_i)$ can be estimated based on Eq. (3-2) while assuming its probability distribution. In the example, the micrometer's resolution from its specification sheet is 1 μm. The probability distribution can be treated as rectangular since the value of a measurand lies in the interval by chance. By applying a half value of the resolution $a=0.5$ μm, $u(Res)=0.5\ \mu m/\sqrt{3}=0.289\ \mu m$.

The standard uncertainty of the measurement repeatability $u(Rpt)$ can be estimated based on the standard deviation σ of repeated independent observations (Type A evaluation). Here, the arithmetic mean and standard deviation of the independent repeated observations were 25.00007 mm and 0.577 μm, respectively. When the arithmetic mean of the three observations ($n=3$) is employed as the measurement result, $u(Rpt)=0.577\ \mu m/\sqrt{3}=0.333\ \mu m$, according to Eq. (3-1). By referring to Eq. (3-3), the sensitivity coefficient $c(IR_t)$ is calculated as in Eq. (3-7).

7)のように算出できる．熱膨張係数には平均値，温度差には最大値である0.5℃を適用している．以上の結果を式(3-4)を用いて合成することで，温度tにおけるマイクロメータの読みIR_tの標準不確かさは$u(IR_t)=\sqrt{u^2(Rpt)+u^2(Res)}=0.441$ μmとなる．

ある文献によると，ブロックゲージの熱膨張係数αは$(11\pm2)\times10^{-6}$/℃である．この値を信頼して，熱膨張係数の不確かさ$u(\alpha_{IR})$を得る（Type B 評価）．確率分布を一様分布とみなし，式(3-2)を適用することで，$a=2\times10^{-6}$，$A=\sqrt{3}$より，熱膨張係数の標準不確かさは$u(\alpha_{IR})=2\times10^{-6}/\sqrt{3}=1.2\times10^{-6}$/℃となる．このときの感度係数$c(\alpha_{IR})$は式(3-3)を用いて式(3-8)のように算出できる．不確かさの過小評価を避けるため，マイクロメータの読みIR_tには最も長いブロックゲージ測定時の平均値を採用した．

$$c(IR_t)=\frac{\partial E_{20}}{\partial IR_t}=1-\alpha_{IR}\Delta t_{IR}=1-11\times10^{-6}\times(20.5-20)=0.99999 \tag{3-7}$$

$$c(\alpha_{IR})=\frac{\partial E_{20}}{\partial \alpha_{IR}}=-IR_t\Delta t_{IR}=-25.0007\times(20.5-20)=-12.5004\ \text{mm℃} \tag{3-8}$$

$$c(\Delta t_{IR})=\frac{\partial E_{20}}{\partial \Delta t_{IR}}=-IR_t\alpha_{IR}=-25.0007\times11\times10^{-6}=-0.00028\ \text{mm/℃} \tag{3-9}$$

$$c(G_{20})=\frac{\partial E_{20}}{\partial G_{20}}=-1 \tag{3-10}$$

$$\begin{aligned} u_c&=\sqrt{\{c(IR_t)u(IR_t)\}^2+\{c(\alpha_{IR})u(\alpha_{IR})\}^2+\{c(\Delta t_{IR})u(\Delta t_{IR})\}^2+\{c(G_{20})u(G_{20})\}^2}\\ &=\sqrt{\{(1-\alpha_{IR}\Delta t_{IR})\cdot u(IR_t)\}^2+\{(IR_t\Delta t_{IR})\cdot u(\alpha_{IR})\}^2+\{(-IR_t\alpha_{IR})\cdot u(\Delta t_{IR})\}^2+\{(-1)\cdot u(G_{20})\}^2}\\ &=0.451 \end{aligned} \tag{3-11}$$

An arithmetic mean value is employed for the thermal expansion coefficient of the block gauge, while the maximum value is employed for temperature t. Combining these uncertainty components by using Eq. (3-4), the standard uncertainty of the reading of micrometer IR_t is calculated as $u(IR_t)=\sqrt{u^2(Rpt)+u^2(Res)}=0.441$ μm.

According to the information in the document, the coefficient of thermal expansion of the block gauge is $(11\pm2)\times10^{-6}$/℃. Relying on this value, a standard uncertainty of the coefficient of thermal expansion $u(\alpha_{IR})$ can be estimated (Type B evaluation). By regarding the probability distribution as rectangular and applying Eq. (3-2) ($a=2\times10^{-6}$, $A=\sqrt{3}$), the standard uncertainty is $u(\alpha_{IR})=2\times10^{-6}/\sqrt{3}=1.2\times10^{-6}$/℃. The sensitivity coefficient $c(\alpha_{IR})$ can also be calculated as shown in Eq. (3-8), based on Eq. (3-3). To avoid underestimation, the arithmetic mean value of the measured length of the longest block gauge (25 mm) was employed.

いま，室温は23.0℃に設定されているが，実際には ±0.5℃の幅でゆるやかに変動している（Type B 評価）．確率分布を一様分布とみなし，$a=0.5$℃，$A=\sqrt{3}$ として，温度差の標準不確かさは $u(\Delta t_{IR})=0.5/\sqrt{3}=0.289$℃となる．なお，このときの感度

Table 3-1 Uncertainty budget table

Source of uncertainty 不確かさ要因	Symbol 記号 x_i	Type タイプ	Uncertainty value 不確かさ値	Probability distribution 確率分布	Divisor 除数	Standard uncertainty 標準不確かさ $u(x_i)$	Sensitivity coefficient 感度係数 c_i	$\lvert c\rvert\times\lvert u\rvert$ μm
Repeatability of reading 測定の繰り返し精度	$u(Rpt)$	A	—	—	—	0.333 μm	1	0.333
Resolution of instruments 測定分解能	$u(Res)$	B	0.001 mm	Rectangular 一様分布	$\sqrt{12}$	0.289 μm	1	0.289
Uncertainty of gauge block ゲージブロックの不確かさ	$u(G_{20})$	B	0.1 μm	Normal 正規分布	2	0.050 μm	−1	0.050
Coefficient of thermal expansion 熱膨張係数	$u(\alpha_{IR})$	B	$\pm2\times10^{-6}$ /℃	Rectangular 一様分布	$\sqrt{3}$	1.154×10^{-6} /℃	−12.5004 mm℃	0.014
Temperature deviation 温度の揺らぎ	$u(\Delta t_{IR})$	B	±0.5 ℃	Rectangular 一様分布	$\sqrt{3}$	0.289 ℃	−0.00028 mm/℃	0.081
Combined uncertainty u_c 合成不確かさ								0.451 μm
Expanded uncertainty U_c（$k=2$, 95% confidence）拡張不確かさ								0.90 μm

Now, the room temperature is 23.0℃. However, it is known that the temperature is slowly changing in a range of ±0.5℃ (Type B evaluation).

係数 $c(\Delta t_{IR})$ は式(3-3)を用いて式(3-9)のように算出できる．熱膨張係数には平均値を適用し，不確かさの過小評価を避けるため，マイクロメータの読み IR_t には最も長いブロックゲージ測定時の平均値を採用した．

ブロックゲージには，校正証明書が添付されている．今回用いたゲージには，「不確かさ 0.10 μm （k=2)」の記載があった（**Type B 評価**）．この不確かさの値は，後に述べる合成不確かさであるため，与えられた値を k の値で除し，標準不確かさに換算して取り扱う（$u(G_{20})$ = 0.1 μm/2 = 0.05 μm）．感度係数は式(3-10)のとおり得られる．

合成不確かさの評価　以上のように求めた各入力量の不確かさと感度係数をもとに，式(3-4)を用いて合成標準不確かさ u_c を式(3-11) のように求める．この u_c に対して包含係数 k を乗ずることで，拡張不確かさ U_c が得られる．k は，厳密には自由度を勘案して決定するが，実用上多くの場合においては，k=2 として扱われ，この場合の信頼水準は 95% となる．

Assuming the probability distribution to be rectangular, a = 0.5℃, $A = \sqrt{3}$, the standard uncertainty of the temperature is $u(\Delta t_{IR}) = 0.5/\sqrt{3}$ = 0.289℃. The sensitivity coefficient $c(\Delta t_{IR})$ can also be calculated as shown in Eq. (3-9), based on Eq. (3-3). The arithmetic mean of the coefficient of thermal expansion was applied to the equation. To avoid underestimation, the arithmetic mean value of the measured length of the longest block gauge (25 mm) was applied.

Each block gauge has a calibration certificate. In the calibration certificate of the block gauge used in the measurement, the following phrase can be found (**Type B evaluation**): "uncertainty: 0.10 μm (k=2)". This value corresponds to the expanded uncertainty, the details of which are described in what follows. Therefore, the standard uncertainty is calculated by dividing the value with k, as $u(G_{20})$ = 0.1 μm/2 = 0.05 μm. The sensitivity coefficient is obtained as shown in Eq. (3-10).

Evaluation of the combined standard uncertainty　By using all the standard uncertainties and the sensitivity coefficients associated with the input values, the combined uncertainty is calculated as shown in Eq. (3-11) based on Eq. (3-4). The expanded uncertainty U_c is obtained as a product of u_c and k. Although the coverage factor k can be determined by considering the degree of freedom, k can be treated as 2 in most cases. In this case, the level of confidence will be approximately 95%.

不確かさバジェット表の作成　これらの算出結果および過程を，Table 3-1 に示すように不確かさバジェット表にまとめる．

【演習問題】

3-1）3.3 節において，繰り返し測定回数を 16 回に増やした場合，標準偏差が一定であるものとして，どの程度まで測定不確かさを低減できるか，検討せよ．

3-2）3.3 節の不確かさ評価を，マイクロメータに代えてノギスで行った場合，どのような不確かさ要素が増加するか，考察せよ．

3-3）測定不確かさ評価における「自由度」について GUM などを参照して調べよ．

Build up the uncertainty budget table　The calculation results and procedure are summarized as the uncertainty budget table shown in Table 3-1.

【Problems】

3-1) Calculate how much you can reduce the measurement uncertainty by increasing the number of replicated measurements to 16 based on the same standard deviation in Section 3.3.

3-2) Consider what kind of error sources will be included when employing a vernier caliper as a measuring instruments for the micrometer in Section 3.3.

3-3) Investigate the "degree of freedom" in the uncertainty evaluation by referring literatures such as the GUM.

第４章　信号の変換

センサで集めた情報は信号処理システムにより処理されて，有用な情報が抽出される．センサ出力，画像信号，音声信号など様々な種類の信号が信号処理システムにより取り扱われる．本章では，計測に用いられるアナログ・デジタル変換，時間域・周波数域変換などについて説明する．

● 4.1　アナログ・デジタル変換

Fig. 4-1 に示すように，信号の振幅が時間に対して測定される場合を考える．振幅と時間の両方が連続なとき，信号はアナログ信号と呼ばれる．これに対して，時間

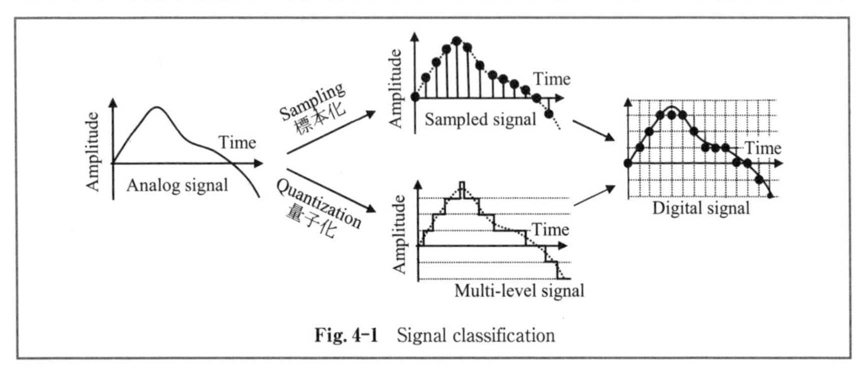

Fig. 4-1　Signal classification

Chapter 4　Signal Conversions

Information collected by sensors is stored and analyzed by signal processing systems, from which useful data are extracted. In signal processing systems, several signals, such as sensor outputs, image signals, sound signals, etc., are processed. In this chapter, analog-digital conversion and time domain-frequency domain conversion are explained for measurements.

● 4.1　Analog-digital conversion

As Fig. 4-1 shows, signal amplitudes are generally measured as a function of time and are classified into analog and digital signals. When both the amplitude and time

と振幅の両方で離散化されている場合，デジタル信号と呼ばれる．また，時間軸のみで離散化されている場合，標本化（サンプル）信号と呼ばれる．振幅のみ離散化されている場合は多値信号または量子化信号と呼ばれる．

アナログ信号は，オペアンプなどで構成されるアナログ回路で処理される．デジタル信号を得るためには，アナログ信号の不要な高周波成分を低域通過フィルタにより除去し，アナログ・デジタル（AD）変換器によりサンプリングと変換が行われる．時間間隔 T_s 秒ごとに信号値を得る．T_s はサンプリング周期，$1/T_s$ はサンプリング周波数と呼ばれる．サンプリングされた信号は，ホールド回路により，その値が保持される．引き続き，AD 変換器は，サンプルした信号を離散的なデジタル値（多くは 2 進符号）に変換する．逆に，デジタル値をアナログ信号に戻す場合は，デジタル・アナログ（DA）変換を行う．

AD 変換器にはいくつかの種類があるが，積分方式と比較方式に分けられる．積分方式はアナログ入力電圧と基準電圧を交互に積分し，電圧に比例したパルス数を計数することでデジタル値を得る方式である．比較方式では，アナログ入力電圧と基準電圧を比較しコンパレータによりデジタル化される．変換速度，精度，コストなどの関

are continuous, the signal is called an "analog signal". On the other hand, when both the amplitude and time are discrete, the signal is called a "digital signal". When only the time is digitized, it is called a "sampled signal". When only the amplitude is digitized, it is called a multi-level signal or quantized signal.

An analog signal is generally processed by analog circuits consisting of operational amplifiers (OP). To obtain a digital signal, the unnecessary high frequency components are removed by a low-pass filter, and the passed signal is sampled and converted by an analog-digital (AD) converter. The signal is sampled with the period of T_s, which is called the "sampling period"; the inverse $1/T_s$ is called the "sampling frequency" or "sample rate". The sampled signal is held by a sampling and hold circuit. Subsequently, the AD converter converts the analog signal into a digital signal (made up mostly of binary numbers). Inversely, a digital signal is converted into analog signal by a digital-analog (DA) converter.

There are several kinds of AD converters. They are generally divided into two categories, integral converters and successive approximation converters. In an integral converter, the analog input voltage and a reference voltage are integrated alternately. By counting the pulses proportional to the voltages, the analog signal is

係で，最も多く利用されている逐次比較型AD変換器について，以下で詳しく説明する．

Fig. 4-2は逐次比較型AD変換器の回路構成を示す．サンプル・ホールド回路により，アナログ入力電圧をある時刻に保持する．この電圧をV_{IN}とし，AD変換器の中に組み込まれたDA変換器からの出力と比較しながら，V_{IN}を最上位(MSB)から最下位(LSB)のNビットのデジタル値に変換する．以下ではFig. 4-2の4ビットAD変換器の場合について，働きを説明する．

DA変換器のスイッチS_0をグランドに接続し，スイッチS_1からS_5を入力電圧V_{IN}に接続すると，S_1, S_2, S_3, S_4, S_5に接続された静電容量$C, C, 2C, 4C, 8C$のコンデンサはすべて電圧V_{IN}による電荷が蓄積される．スイッチ$S_0, \ldots, S_5$を切り離すとV_{IN}により蓄積された電荷はそのまま保持される．その後，スイッチS_1, S_2, S_3, S_4, S_5をグラ

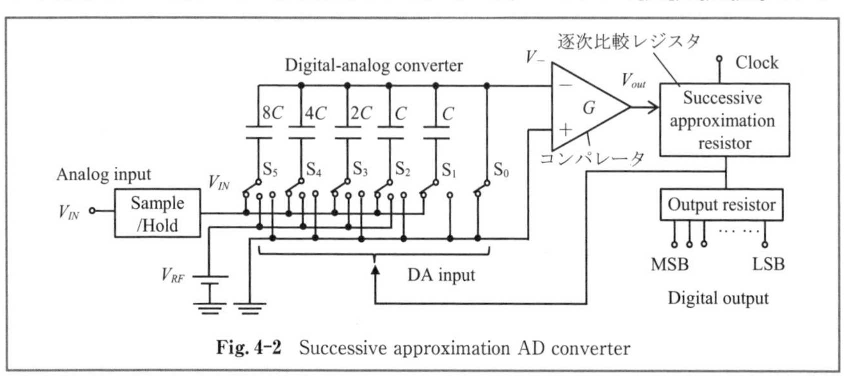

Fig. 4-2 Successive approximation AD converter

converted into a digital signal. In a successive approximation converter, the analog input voltage is compared with reference voltages, and is successively converted by a comparator into a digital value. For reasons of speed, precision, and cost of conversion, the successive approximation converter is used most often. The mechanism of a successive approximation AD converter is described below.

Fig. 4-2 shows the circuit of a successive approximation converter. First, an analog voltage V_{IN} is captured and held for a certain time by sampling and hold circuits. Comparing V_{IN} with an output of the DA converter installed in the AD converter, the analog value of V_{IN} is converted into a digital value composed of N bits from the most significant bit (MSB) to the least significant bit (LSB). In what follows, the conversion operation is explained using Fig. 4-2, which shows a 4-bit AD converter.

First, switch S_0 of the DA converter is grounded and the switches from S_1 to S_5 are connected to the input voltage V_{IN}. Then, the capacitors C, C, $2C$, $4C$, and $8C$ are

ンドに接続すると，コンパレータの負端子（コンデンサの共通端子）の電圧 V_- は $-V_{IN}$ に電位が下げられる．ここまでが比較のための準備である．

(1) 比較においては，まずMSBを決定するために V_{IN} と参照電圧 V_{RF} の1/2と比較する．このためDA変換器には $V_{RF}/2$ のデジタル値1000が入力される．すなわち，スイッチ S_5 の端子の電圧はグランドから参照電圧 V_{RF} に切り替えられる．このとき電圧 V_- は，コンデンサの容量比 $8C:4C+2C+C+C=1:1$ より，$V_{RF}/2$ が加算されて，$V_-=-V_{IN}+V_{RF}/2$ となる．ここで，たとえば，$V_{IN}=0.7\,V_{RF}$ と仮定する．V_{IN} が $V_{RF}/2$（2進数1000）より大きいので，V_-（$=-0.7\,V_{RF}+V_{RF}/2$）は負となりコンパレータの出力端子 V_{out} は正（デジタル値1，$V_{out}=1$）となる．すなわち，出力レジスタのMSB=1となる．

(2) 次にMSBより1桁下を決めるため，デジタル値1100がDA変換器に入力され，スイッチ S_5 に加えて S_4 も V_{RF} に接続される．このとき $8C$ と $4C$ のコンデンサが並列で V_{RF} に接続され，他のコンデンサはグランドに接続されたままであるので，V_- はコンデンサの容量比 $8C+4C:2C+C+C=3:1$ より，$V_-=-0.7\,V_{RF}+(3/4)V_{RF}>0$ となり $V_{out}=0$ となる．したがって第2ビットは0となる．

connected to switches S_1, S_2, S_3, S_4, and S_5, which are all charged at the voltage V_{IN}. When switches S_0, S_1, S_2, S_3, S_4, and S_5 are disconnected, the accumulated charges by V_{IN} are held without discharging. Then, switches S_1, S_2, S_3, S_4, and S_5 are grounded, the voltage V_- of the negative terminal of the comparator (the common terminal of condensers) is lowered to be $-V_{IN}$. The procedure described above is the preparation for approximation.

(1) In the approximation, first, in order to determine MSB, V_{IN} is compared with one half of the reference voltage V_{RF}. For this purpose, the digital value of 1000 corresponding to $V_{RF}/2$ is set in the DA converter; i.e. S_5 is switched from the ground to V_{RF}. Then, the voltage V_- is increased to be $V_-=-V_{IN}+V_{RF}/2$ by adding $V_{RF}/2$, which is determined by the ratio of capacitance $8C:4C+2C+C+C=1:1$. Here, as an example, we assume $V_{IN}=0.7V_{RF}$. As V_{IN} is larger than $V_{RF}/2$ (with a digital value 1000), the voltage V_- ($=-0.7V_{RF}+V_{RF}/2$) is negative and the output V_{out} becomes positive (digital value of 1, i.e. $V_{out}=1$). Thus, MSB=1.

(2) In order to determine the digital value one order lower than MSB, the digital value in the DA converter is set to 1100, the terminal of S_4 is connected to V_{RF} as well as S_5. Then, condensers $8C$ and $4C$ are connected to V_{RF} in parallel. Since the other

(3) 次にデジタル値1010がDA変換されるよう，S_5, S_3がV_{RF}に接続され，S_4, S_2がグランドに接続される．コンデンサの容量比は$8\,C+2\,C:4\,C+C+C=5:3$となり，$V_-=-0.7\,V_{RF}+(5/8)\,V_{RF}<0$であるので，$V_{out}=1$となる．

(4) 最後にデジタル値1011が入力され，S_5, S_3, S_2がV_{RF}に接続され，S_4はグランドに設定される．コンデンサの容量比は11：5となり，$V_-=-0.7\,V_{RF}+(11/16)\,V_{RF}<0$であるので，$V_{out}=1$となる．これによりデジタル出力値は1011となる．1011に対応する電圧は$0.6875\,V_{RF}$であるので，入力電圧$0.7\,V_{RF}$に近い電圧のデジタル値になっている．

● 4.2　時間域・周波数域変換

信号の時間波形の一般的な計測では，オシロスコープがしばしば用いられる．観測

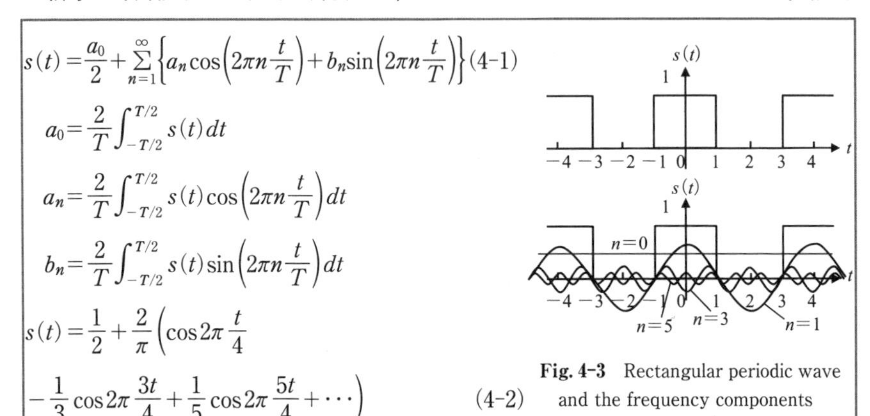

$$s(t)=\frac{a_0}{2}+\sum_{n=1}^{\infty}\left\{a_n\cos\left(2\pi n\frac{t}{T}\right)+b_n\sin\left(2\pi n\frac{t}{T}\right)\right\}\quad(4\text{-}1)$$

$$a_0=\frac{2}{T}\int_{-T/2}^{T/2}s(t)\,dt$$

$$a_n=\frac{2}{T}\int_{-T/2}^{T/2}s(t)\cos\left(2\pi n\frac{t}{T}\right)dt$$

$$b_n=\frac{2}{T}\int_{-T/2}^{T/2}s(t)\sin\left(2\pi n\frac{t}{T}\right)dt$$

$$s(t)=\frac{1}{2}+\frac{2}{\pi}\left(\cos 2\pi\frac{t}{4}-\frac{1}{3}\cos 2\pi\frac{3t}{4}+\frac{1}{5}\cos 2\pi\frac{5t}{4}+\cdots\right)\quad(4\text{-}2)$$

Fig. 4-3　Rectangular periodic wave and the frequency components

condensers are still grounded, V_- becomes $V_-=-0.7\,V_{RF}+(3/4)\,V_{RF}>0$ due to the ratio of $8C+4C:2C+C+C=3:1$. Therefore, $V_{out}=0$, and the second bit is equal to 0.

(3) Next, in order to convert the digital value of 1010, S_5 and S_3 are connected to V_{RF}, and S_4 and S_2 are grounded. The ratio of the capacitance is now $8C+2C:4C+C+C=5:3$, and $V_-=-0.7\,V_{RF}+(5/8)\,V_{RF}<0$. Thus, $V_{out}=1$.

(4) Finally, a digital value of 1011 is set, S_5, S_3, and S_2 are connected to V_{RF}, and S_4 is grounded. The ratio of the condensers is now 11：5 and $V_-=-0.7\,V_{RF}+(11/16)\,V_{RF}<0$. Thus, $V_{out}=1$. Therefore, the digital output is 1011, which is corresponding to $0.6875\,V_{RF}$, the approximated value is close to the input value of $0.7\,V_{RF}$.

● 4.2　Time domain-frequency domain conversion

In conventional measurement of signal waveform, an oscilloscope is often used. It

した波形の中の信号の周波数およびノイズ周波数を知ることは重要である．信号を周波数領域で調べるためには，時間領域の波形をフーリエ変換し，信号波形を構成する周波数成分（スペクトル）を知ることが必要である．ここで，周期がTである信号$s(t)$のフーリエ級数表現を考える．成分として基本波周波数$f(=1/T)$とその整数倍$2f(=2/T)$，$3f(=3/T)$,...,を考えることができる．さらに，直流成分$a_0/2$を加え，周期信号$s(t)$は式(4-1)により表現される．三角関数の直交性より各成分の係数を計算できる．例としてFig. 4-3のような矩形周期波形の場合は，$T=4$秒であり，式(4-2)で表される．それぞれの余弦成分a_nは$n=1, 3, 5, \ldots$に対して，波形は$f, 3f, 5f$, ...の周波数を持つ成分で構成され，それらは，$2/\pi(=0.637)$，$-2/(3\pi)(=-0.212)$，$2/(5\pi)(=0.127)\cdots$となる．また$a_0=1$，$b_n=0(n=1, 2, 3, \ldots)$である．Fig. 4-3にそ

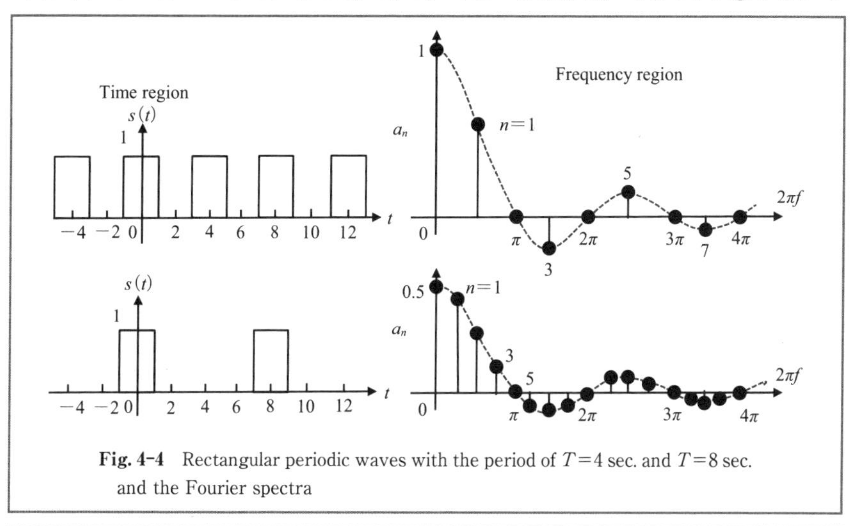

Fig. 4-4　Rectangular periodic waves with the period of $T=4$ sec. and $T=8$ sec. and the Fourier spectra

is important to know the signal and noise frequencies in the measured waveform. In order to examine the signal in the frequency domain, the waveform in the time domain is Fourier transformed and the signal frequency components (i.e., the spectrum) forming the signal waveform are examined. Here, we consider the Fourier series expression of signal $s(t)$ with a period of T. Then, the component at the fundamental frequency f $(=1/T)$ and those at harmonic frequencies $2f(=2/T)$, $3f$ $(=3/T)$,... are considered. Adding the direct current (DC) component $a_0/2$, the periodic signal $s(t)$ is expressed by Eq. (4-1). Each coefficient of the component is calculated by the orthogonality of sinusoidal functions. As an example, the rectangular periodic waveform shown in Fig. 4-3 is expressed by Eq. (4-2) with the period of $T=4$. For the cosine components a_n $(n=1, 3, 5, \ldots)$, the waveform consists of the components

れぞれの周波数成分の時間波形を示した．各成分の和により，元の波形が構成される．Fig. 4-4 に周期 $T=4$ 秒および $T=8$ 秒の矩形周期波形とフーリエ級数の係数 a_n ($n=1, 2, 3, \ldots$) を示した．スペクトルは離散的となる．周期が増えるに従い，周波数領域の離散的なスペクトルは密に分布する．

信号の処理として一般にフィルタが用いられるが，周波数領域のスペクトル分布を考えるとフィルタの働きが分かりやすい．例えば低域通過フィルタにより，高周波成分を取り除く場合について，Fig. 4-3，Fig. 4-4 で求めた時間波形とスペクトルを用いて考える．Fig. 4-5 に示すような低域フィルタを RC 回路により考えるとき，フィルタのカットオフ周波数($f_c=1/(2\pi RC)$)を 3/8 Hz に設定すると，矩形波の基本周波数成分は，離散的スペクトルの $n=1$ (a_1) で示され，周波数は $f=1/4$ Hz であるので，$n=1$ のスペクトルだけ取り出せる．出力波形は a_0 と a_1 で示される直流オフセットの

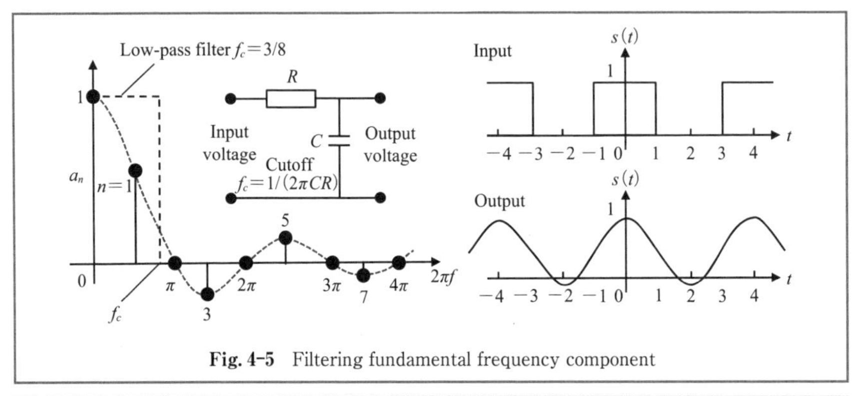

Fig. 4-5 Filtering fundamental frequency component

with frequencies of f, $3f$, $5f$, ..., and are given by $2/\pi$ (=0.637), $-2/(3\pi)$ (=−0.212), $2/(5\pi)$ (=0.127) respectively, while $a_0=1$, $b_n=0$ ($n=1, 2, 3, \ldots$). Fig. 4-3 shows the waveform of the respective frequency components. Adding the respective components, the original waveform is reconstructed. Fig. 4-4 shows the coefficients a_n ($n=1, 2, 3, \ldots$) of Fourier series for the periods $T=4$ s and $T=8$ s. The spectrum is discrete. When the period is increased, the spectrum becomes dense in a frequency domain.

For signal processing, a filter is often used. Filtering is easily understood by considering the spectrum in the frequency domain. As an example, in order to remove high-frequency noise, consider a low-pass filter in the case of the waveform shown in Fig. 4-3 and the spectrum in Fig. 4-4. If the low-pass filter of the RC circuit shown in the inset of Fig. 4-5 has a cutoff frequency ($f_c=1/(2\pi RC)$) of 3/8 Hz, the spectrum for $n=1$ (a_1) is passed since the fundamental frequency of the rectangular waveform

ある基本周波数の正弦波となる．

正弦と余弦関数を用いて表されたフーリエ級数をオイラーの公式により複素数で表現すると，式(4-3)で定義される複素フーリエ級数が得られる．フーリエ級数では無限に続く周期波形のみしか扱えない．単一パルスのような周期波形でない場合を扱えるようにするため，周期Tを大きくすることを考える．式(4-3)において，$T\to\infty$を考え，$\Delta f=1/T$，$n\Delta f=f$，Σを$\int$に置き換え，$c_nT=S(f)$とすると，級数表現は式(4-4)で表される積分表現となる．$S(f)$をフーリエ変換と呼び，元の時間関数$s(t)$はフーリエ逆変換により得られる．Fig. 4-6に矩形パルス波形とそのフーリエ変換を示す．Tを大きくすると離散的スペクトルの周波数の間隔が小さくなり，連続スペ

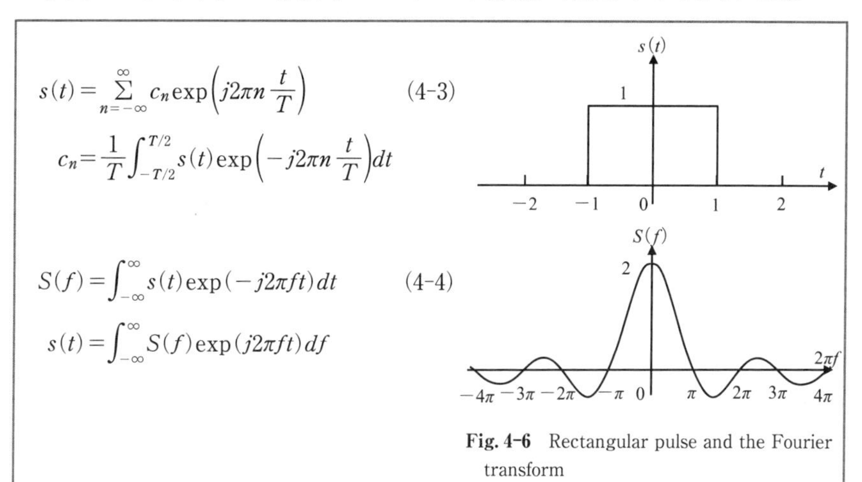

$$s(t)=\sum_{n=-\infty}^{\infty}c_n\exp\left(j2\pi n\frac{t}{T}\right) \tag{4-3}$$

$$c_n=\frac{1}{T}\int_{-T/2}^{T/2}s(t)\exp\left(-j2\pi n\frac{t}{T}\right)dt$$

$$S(f)=\int_{-\infty}^{\infty}s(t)\exp(-j2\pi ft)\,dt \tag{4-4}$$

$$s(t)=\int_{-\infty}^{\infty}S(f)\exp(j2\pi ft)\,df$$

Fig. 4-6　Rectangular pulse and the Fourier transform

is equal to 1/4 Hz.

Fourier series expressed in terms of sine and cosine functions are converted to complex Fourier series by Euler's formula, which is defined by Eq. (4-3). Fourier series only express periodic waveforms lasting infinitely in time. In order to express a single-pulse waveform that is not periodic, consider that the period T becomes infinitely long. In Eq. (4-3), $T\to\infty$, and replacing as $\Delta f=1/T$, $n\Delta f=f$, $\Sigma\to\int$ and $c_nT=S(f)$, the series expression is converted to the integral expression shown in Eq. (4-4). $S(f)$ is called Fourier transform and the original time signal $s(t)$ is given by the inverse Fourier transform. In Fig. 4-6, a single rectangular pulse and its Fourier transform are shown. By increasing the period T, the frequency interval of the discrete spectrum becomes small, and the spectrum finally becomes continuous. The

クトルとなる．Fig. 4-4 に示した離散的スペクトルの包絡線が単一パルスの連続スペクトルとして得られる．

● 4.3 サンプリング定理と離散フーリエ変換

AD 変換において時間域のサンプリング周期をどのように選ぶかは重要な問題である．波形の変化より遅いサンプリング周期では，波形を正しく計測できない．波形の変化より十分速いサンプリング周期であれば，元の波形を再生できるが，冗長な信号となる．サンプリング定理は，元の信号を復元するのに必要な必要最小限のサンプリング周波数を与える．

サンプリング定理（標本化定理） 信号に含まれる最大周波数が f_M であるとき，$2f_M$ 以上のサンプリング周波数でサンプリングを行うとき，元の信号を復元することができる．例えば，コンパクトディスクの場合，音声帯域が 20 Hz～20 kHz であるが，サンプリング周波数として 44.1 kHz が用いられている．

$$s(t)=\int_{-\infty}^{\infty}S(f)\exp(j2\pi ft)df \quad (4\text{-}5) \qquad s(t)=\int_{-f_M}^{f_M}S(f)\exp(j2\pi ft)df \quad (4\text{-}6)$$

$$S(f)=\sum_{n=-\infty}^{\infty}C_n\exp\left(-j2\pi n\frac{f}{2f_M}\right) \quad C_n=\frac{1}{2f_M}\int_{-f_M}^{f_M}S(f)\exp\left(j2\pi n\frac{f}{2f_M}\right)df \quad (4\text{-}7)$$

$$C_n=\frac{1}{2f_M}\int_{-f_M}^{f_M}S(f)\exp\left(j2\pi n\frac{f}{2f_M}\right)df=\frac{1}{2f_M}s\left(\frac{n}{2f_M}\right) \quad (4\text{-}8)$$

envelope of the discrete spectrum shown in Fig. 4-4 corresponds to the continuous spectrum of the pulse.

● 4.3 Sampling theorem and discrete Fourier transform

It is important to determine the sample rate for the AD converter. If the sample rate is too slow to capture the change rate of waveform, the correct waveform is not measured. On the other hand, if the sample rate is too fast, the time signal is reconstructed but the rate is redundant. The sampling theorem gives a sample rate that is sufficient for perfect fidelity in the reconstruction of the sampled signal.

Sampling theorem When the maximum frequency included in a signal is f_M, the original signal is perfectly reconstructed at a sample rate higher than $2f_M$. For example, the voice band is from 20 Hz to 20 kHz, so the sample rate of a compact disc is 44.1 kHz.

以下にサンプリング定理を証明する．時間域の信号$s(t)$は，周波数領域のスペクトル$S(f)$を用いて，逆フーリエ変換の式(4-5)により表される．ここでf_Mを超える周波数成分がないので式(4-5)は式(4-6)となる．式(4-6)より$S(f)$は周波数が$-f_M<f<f_M$の範囲で定義されている関数と考えられるので，$S(f)$について，周波数fの全領域では$-f_M<f<f_M$の範囲の関数が周期$2f_M$で繰り返されていると考える．したがって，$S(f)$はfに関してフーリエ級数の式(4-7)で表せる．fに関する周期は，$2f_M$となり，複素フーリエ級数の係数は，式(4-6)より，式(4-8)となる．すなわち，$s(t)$の$1/(2f_M)$ごとのサンプル$s(n/(2f_M))$ $(n=0, \pm1, \pm2, \ldots)$から元の波形のフーリエ係数が求められ，スペクトル$S(f)$が確定できる．$S(f)$が求められれば，$s(t)$は逆フーリエ変換により決定できる．したがって，サンプリング定理が証明された．

$$s_{smp}(t)=\sum_{n=0}^{\infty} s(n)\delta(t-nT_s) \tag{4-9}$$

$$\begin{aligned}S_{smp}(f)&=\int_{-\infty}^{\infty}\sum_{n=0}^{\infty} s(n)\delta(t-nT_s)\exp(-j2\pi ft)\,dt\\&=\sum_{n=0}^{\infty} s(n)\int_{-\infty}^{\infty}\delta(t-nT_s)\exp(-j2\pi ft)\,dt\\&=\sum_{n=0}^{\infty} s(n)\exp(-j2\pi fnT_s)\end{aligned} \tag{4-10}$$

$$S_{smp}(f)=S(k)=\sum_{n=0}^{N-1} s(n)\exp\left(-j\frac{2\pi}{N}kn\right)=\sum_{n=0}^{N-1} s(n)\left\{\exp\left(-j\frac{2\pi}{N}\right)\right\}^{kn} \tag{4-11}$$

$$s(n)=\frac{1}{N}\sum_{k=0}^{N-1} S(k)\exp\left(j\frac{2\pi}{N}kn\right)=\frac{1}{N}\sum_{k=0}^{N-1} S(k)\left\{\exp\left(-j\frac{2\pi}{N}\right)\right\}^{-kn} \tag{4-12}$$

$$\begin{aligned}S(k)&=\sum_{n=0}^{N-1}\sin(2\pi fnT_s)\exp\left(-j\frac{2\pi}{N}kn\right)\\&=\frac{1}{2j}\sum_{n=0}^{N-1}\left[\exp\left\{j2\pi n\left(fT_s-\frac{k}{N}\right)\right\}-\exp\left\{-j2\pi n\left(fT_s+\frac{k}{N}\right)\right\}\right]\\&=\frac{1}{2j}\sum_{n=0}^{N-1}\left[\exp\left\{j2\pi n\left(\frac{1}{8}-\frac{k}{8}\right)\right\}-\exp\left\{-j2\pi n\left(\frac{1}{8}+\frac{k}{8}\right)\right\}\right]\end{aligned} \tag{4-13}$$

The sampling theorem is proven as follows. The time domain signal $s(t)$ is expressed by the inverse Fourier transform of Eq. (4-5), using the frequency domain signal $S(f)$. Here, the frequency range is limited within f_M, and Eq. (4-5) is modified to become Eq. (4-6). Since $S(f)$ is considered to be defined within the frequency range $-f_M<f<f_M$ in Eq. (4-6), it is consistent to extend the function $S(f)$ defined within $-f_M<f<f_M$ to be repeated periodically in the entire frequency domain with a period of $2f_M$. Thus, $S(f)$ is expressed by Fourier series, as in Eq. (4-7). The period in terms of f is equal to $2f_M$ and the complex coefficients of the Fourier series become

次にサンプル値のフーリエ変換について考える．サンプリング周期を T_s として，デジタルのサンプル値 $s(nT_s)$ （$=s(n)$ $n=0, 1, 2, 3, \ldots$）に変換される．サンプル値全体 $s_{smp}(t)$ は式(4-9)で与えられ，サンプル値のフーリエ変換 $S_{smp}(f)$ は式(4-10)で与えられる．ここで一般には，サンプリング数 N は有限であるので，$s(0), s(1), s(2), \ldots, s(N-1)$ の N 個の点からなるとする．N 個の信号点全部で一周期になるような周期波形を考える．この周期に対する周波数は $1/(NT_s)$ となり，基本周波数とする．このとき基本周波数の整数倍の周波数は $f=k/(NT_s)$ （$k=1, 2, 3, \ldots$）と表せる．したがって，式(4-11)となる．式(4-11)を離散フーリエ変換（DFT）と呼ぶ．周波数 f は離散的な周波数に関する変数 k に置き換えられている．また，逆離散フーリエ変換は，式(4-12)で与えられる．

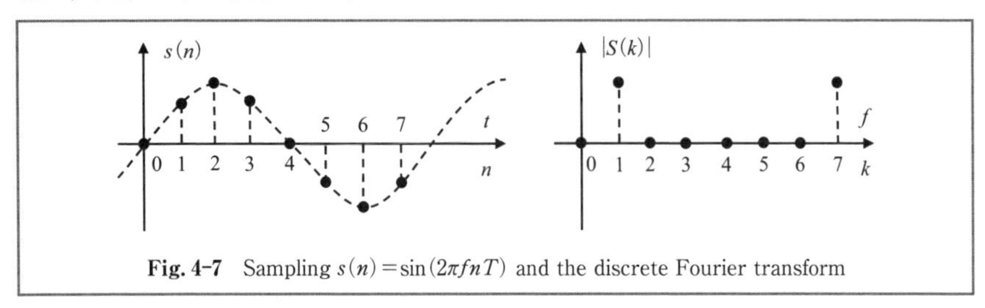

Fig. 4-7　Sampling $s(n)=\sin(2\pi fnT)$ and the discrete Fourier transform

Eq. (4-8) from Eq. (4-6). That is, the Fourier series coefficients are obtained from the sampled values $s(n/(2f_M))$ ($n=0, \pm1, \pm2, \ldots$) of the original time domain signal $s(t)$ at the sampling period $1/(2f_M)$, and the spectrum $S(f)$ is decided, while $s(t)$ is determined by the inverse Fourier transform. Therefore, the sampling theorem is verified.

Next, let us consider the Fourier transform of sampled signals. At the sampling period T_s, the signal is converted into sampled values $s(nT_s)$ ($=s(n)$ $n=0, 1, 2, 3, \ldots$). The sampled signal is completely expressed by Eq. (4-9) and the Fourier transform of the signal is given by Eq. (4-10). Here, as sample number N is generally finite, the signal consists of the N signal points $s(0), s(1), s(2), \ldots, s(N-1)$. Consider a periodic waveform in which one period wave is made up of N signal points. The frequency corresponding to the period is $1/(NT_s)$, this is the fundamental frequency of the signal. The harmonics of the fundamental frequency are expressed by $f=k/(NT_s)$ ($k=1, 2, 3, \ldots$), and Eq. (4-11) is obtained. Eq. (4-11) is called a discrete Fourier transform (DFT), where the frequency f is replaced by the variable k. The inverse Fourier transform is given by Eq. (4-12). As an example, let us obtain the Fourier transform of $s(n)=\sin(2\pi fnT_s)$, where $f=10$ kHz, the sample rate is 80 kHz, and N

例として，$s(n)=\sin(2\pi fnT_s)$の DFT を求める．ここで $f=10$ kHz，サンプリング周波数 80 kHz，$N=8$ とする．このとき式(4-13)が得られる．$k=0, 1, 2, \ldots, 7$ について計算すると，$S(1)=-4j$，$S(7)=4j$，他の $S(k)$ は 0 となる．したがって離散フーリエ変換の大きさ $|S(k)|$ を考えると，Fig. 4-7 に示すように，$k=1$ と $k=7$ に大きさ 4 の値を持つ周波数スペクトルとなる．$k=1$ に対応する周波数は 10 kHz であるので，基本周波数である．$k=7$ は 70 kHz の正弦波に対応するが，実際には存在しない周波数である．この周波数の正弦波のサンプリング値は $k=1$ の場合と同じになり，この周波数成分があるように見え，この現象はエイリアシングと呼ばれる．

デジタル信号の周波数域フィルタを実現するためには，ソフトウェアのプログラムでデジタル信号を処理する．Fig. 4-5 に示した CR フィルタをデジタル信号に適用するには，CR 回路の応答を表す方程式を差分方程式に書き直し，サンプリング値を逐次処理する．これはフィルタの伝達関数をソフトウェアとして実現することである．伝達関数はラプラス変換で表されるが，サンプリング値のラプラス変換は z 変換であるので，これらのデジタル信号処理は z 変換に基づいて行われる．

$=8$. Under these conditions, Eq. (4-13) is obtained. Calculating the values of DFT, $S(1)=-4j$, $S(7)=4j$, and other values of $S(k)$ are 0. Therefore, the magnitude $|S(k)|$ of the DFT is plotted as shown in Fig. 4-7, where, the values are 4 for $k=1$ and 7. The frequency for $k=1$ corresponds to 10 kHz, and is thus the fundamental frequency. For $k=7$, the frequency is 70 kHz, which does not exist in the original signal. The sampled values of the sinusoidal function with frequency 70 kHz are exactly the same as those for $k=1$, and this frequency is a false frequency. This phenomenon is called "aliasing".

In order to realize a frequency filter for a digital signal, the digital signal is processed by software. To apply the CR filter shown in Fig. 4-5 to a digital signal, the time dependent equation of the CR circuit response is converted to a differential equation, and the sampled signals are processed sequentially. This corresponds to the realization of a filter transfer function carried out by the software. The transfer function of a system is generally expressed by using a Laplace transform. The Laplace transform of sampled signals is a z-transform, so the digital signal processing is theoretically expressed on the basis of the z-transform.

【演習問題】

4-1) 8ビットの逐次比較型AD変換器の構成を述べよ．

4-2) Fig. 4-6のパルス波形のフーリエ変換を求めよ．

4-3) Fig. 4-8に示す$s(n)=1(0\leqq n\leqq 4)$，$s(n)=0(5\leqq n\leqq 7)$の波形のDFTを求めよ．ただし$N=8$とする．

4-4) $s(n)=\cos(4\pi fnT_s)$のDFTを求めよ．ここで$f=10$ kHz，サンプリング周波数80 kHz，$N=8$とする．

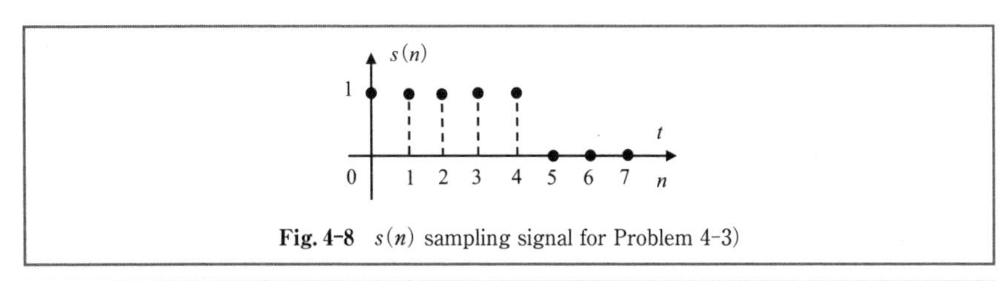

Fig. 4-8　$s(n)$ sampling signal for Problem 4-3)

【Problems】

4-1) Show the construction of an 8-bit successive approximation AD converter.

4-2) Find the Fourier transform of the rectangular single pulse shown in Fig. 4-6.

4-3) Find the DFT of $s(n)=1$ $(0\leqq n\leqq 4)$, and $s(n)=0$ $(5\leqq n\leqq 7)$ shown in Fig. 4-8, where $N=8$.

4-4) Find the DFT of $s(n)=\cos(4\pi fnT_s)$, where $f=10$ kHz, the sample rate is 80 kHz, and $N=8$.

第5章　データ処理

計測した諸量を工学に役立てるためには，相互の諸量の相関関係や時系列変化の把握が必要であり，そのためには計測した結果をデータ処理する必要があるので，本章では，基本的なデータ処理について述べる．

● 5.1　データの統計処理

2つの物理量を幾組か計測して相互の変数間の関係を，計測値と近似曲線との差異の二乗和が最小となる最小二乗法で求める方法について述べる．一例として，Fig. 5-1のように，2つの変数 x と y に線形関係 $y=ax+b$ が成り立つ場合について考える．この場合，N 組の計測値 (x_i, y_i) $(i=1, 2, 3, \dots, N)$ の計測結果から式(5-2)が得られ

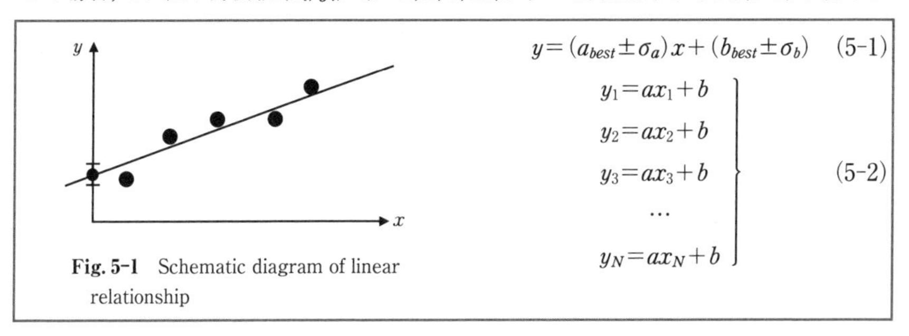

Fig. 5-1　Schematic diagram of linear relationship

$$y=(a_{best}\pm\sigma_a)x+(b_{best}\pm\sigma_b) \quad (5\text{-}1)$$

$$\left.\begin{aligned} y_1&=ax_1+b \\ y_2&=ax_2+b \\ y_3&=ax_3+b \\ &\cdots \\ y_N&=ax_N+b \end{aligned}\right\} \quad (5\text{-}2)$$

Chapter 5　Data Processing

In order to apply measured values to engineering, classifications of the relationships between measured values and processing of time-series data are required. Thus, data processing is discussed in this chapter.

● 5.1　Statistical processing of data

This section explains how to obtain an approximate curve by the least squares method. In the least squares method, summation of the squares of the difference between the measured values and the approximate curve has a minimum. Initially, a linear relationship of $y=ax+b$ is considered (Fig. 5-1). When N sets of data (x_i, y_i) $(i=1, 2,$

る．式(5-2)を観測式という．未知定数の数と観測式の数が同数であれば連立方程式により未知定数が一意に求まる．未知定数の数より多いN組の計測値の場合は，以下のようにして最良推定値a_{best}，b_{best}とそれぞれの誤差σ_a，σ_bを最小二乗法により求めることができる（式(5-1)参照）．各計測値y_iが幅σ_yの正規分布に従っていると仮定すれば，y_iを観測する確率$P_{a,b}(y_i)$は式(5-3)である．$y_1, \ldots, y_N$の測定値を観測する確率は式(5-4)，(5-5)で表される．確率$P_{a,b}(y_1, \ldots, y_N)$を最大にする$a$と$b$は二乗

$$P_{a,b}(y_i) \propto \frac{1}{\sigma_y} e^{-\frac{(y_i - ax_i - b)^2}{2\sigma_y^2}} \tag{5-3}$$

$$P_{a,b}(y_1, \ldots, y_N) = P_{a,b}(y_1) \cdots P_{a,b}(y_N) \propto \frac{1}{\sigma_y{}^N} e^{-\frac{\chi^2}{2}} \tag{5-4}$$

$$\chi^2 = \sum_{i=1}^{N} \frac{(y_i - ax_i - b)^2}{\sigma_y{}^2} \tag{5-5}$$

$$\frac{\partial \chi^2}{\partial a} = \frac{-2}{\sigma_y{}^2} \sum_{i=1}^{N} x_i (y_i - ax_i - b) \tag{5-6}$$

$$\frac{\partial \chi^2}{\partial b} = \frac{-2}{\sigma_y{}^2} \sum_{i=1}^{N} (y_i - ax_i - b) \tag{5-7}$$

$$a_{best} = \frac{N \sum_{i=1}^{N} x_i y_i - \sum_{i=1}^{N} x_i \sum_{i=1}^{N} y_i}{\Delta} \tag{5-8}$$

$$b_{best} = \frac{\sum_{i=1}^{N} x_i{}^2 \sum_{i=1}^{N} y_i - \sum_{i=1}^{N} x_i \sum_{i=1}^{N} x_i y_i}{\Delta} \tag{5-9}$$

$$\Delta = N \sum_{i=1}^{N} x_i{}^2 - \left(\sum_{i=1}^{N} x_i \right)^2 \tag{5-10}$$

$$\sigma_a = \sigma_y \sqrt{\frac{N}{\Delta}} \tag{5-11}$$

$$\sigma_b = \sigma_y \sqrt{\frac{\sum_{i=1}^{N} x_i{}^2}{\Delta}} \tag{5-12}$$

$$\sigma_y = \sqrt{\frac{1}{N-2} \sum_{i=1}^{N} \{ y_i - (a_{best} x_i + b_{best}) \}^2} \tag{5-13}$$

3, …, N) are used, Eq. (5-2) is obtained. Eq. (5-2) is called an observation equation. If the number of unknown parameters equals the number of observation equations, the unknown parameters are uniquely obtained. If the number of sets of measured values N is larger than that of unknown parameters, the best estimated values a_{best} and b_{best} and their errors σ_a and σ_b are obtained by the least squares method (see Eq. (5-1)). When it is assumed that each measured value y_i follows a normal distribution with width σ_y, probability of y_i, $P_{a,b}(y_i)$, is expressed by Eq. (5-3). Probability of y_1, …,

和 χ^2 を最小にする値である．この a と b は，式(5-6)，(5-7)のように χ^2 を a と b でそれぞれ微分して得られた導関数を 0 とし，連立一次方程式を立てて求めることができる．式(5-8)，(5-9)に a と b の最良推定値 a_{best}，b_{best} を示す．それぞれの誤差 σ_a，σ_b は式(5-11)，(5-12)で求めることができる．なお，式(5-8)，(5-9)は，式(5-2)の両辺に a の係数を掛けて総和を取った a の正規式である式(5-14)と，式(5-2)の両辺に b の係数を掛けて総和を取った b の正規式である式(5-15)から a と b についての連立一次方程式として整理する，と表現することもできる．

Fig. 5-2 に示すように変数 x と y に原点を通る線形関係 $y=ax$ が成り立つ場合には，式(5-16)の最良推定値 a_{best} とその誤差 σ_a は式(5-17)，(5-18)で求められる．なお，

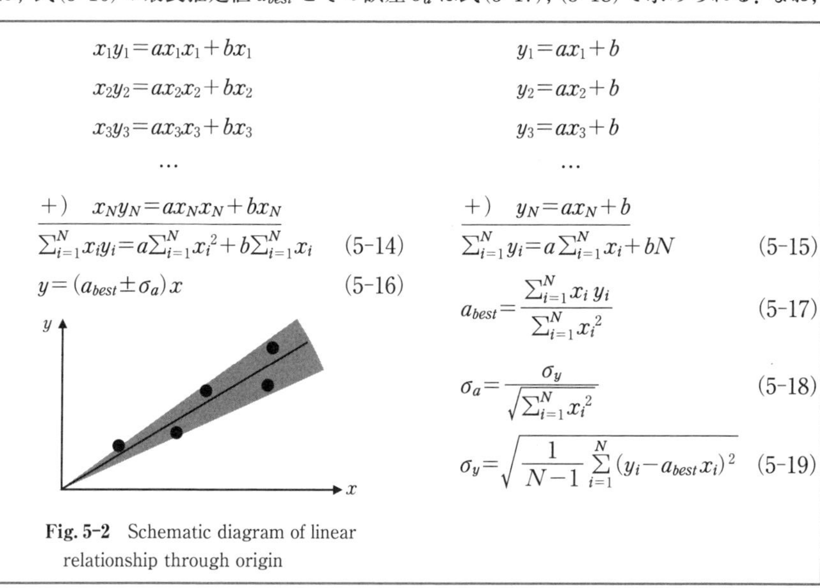

$$
\begin{aligned}
x_1y_1 &= ax_1x_1 + bx_1 \\
x_2y_2 &= ax_2x_2 + bx_2 \\
x_3y_3 &= ax_3x_3 + bx_3 \\
&\cdots \\
+)\quad x_Ny_N &= ax_Nx_N + bx_N
\end{aligned}
$$

$$\textstyle\sum_{i=1}^{N} x_iy_i = a\sum_{i=1}^{N} x_i^2 + b\sum_{i=1}^{N} x_i \qquad (5\text{-}14)$$

$$
\begin{aligned}
y_1 &= ax_1 + b \\
y_2 &= ax_2 + b \\
y_3 &= ax_3 + b \\
&\cdots \\
+)\quad y_N &= ax_N + b
\end{aligned}
$$

$$\textstyle\sum_{i=1}^{N} y_i = a\sum_{i=1}^{N} x_i + bN \qquad (5\text{-}15)$$

$$y = (a_{best} \pm \sigma_a)x \qquad (5\text{-}16)$$

$$a_{best} = \frac{\sum_{i=1}^{N} x_i\, y_i}{\sum_{i=1}^{N} x_i^2} \qquad (5\text{-}17)$$

$$\sigma_a = \frac{\sigma_y}{\sqrt{\sum_{i=1}^{N} x_i^2}} \qquad (5\text{-}18)$$

$$\sigma_y = \sqrt{\frac{1}{N-1}\sum_{i=1}^{N}(y_i - a_{best}x_i)^2} \qquad (5\text{-}19)$$

Fig. 5-2　Schematic diagram of linear relationship through origin

y_N is revealed by Eqs. (5-4) and (5-5). The maximizing of $P_{a,b}$ $(y_1, \ldots, y_N)$ by a and b means minimizing of χ^2. These a and b are obtained from Eqs. (5-6) and (5-7). Eqs. (5-8) and (5-9) show a_{best} and b_{best}. The errors σ_a and σ_b are obtained by Eqs. (5-11) and (5-12) respectively. Eqs. (5-8) and (5-9) are obtained from Eqs. (5-14) and (5-15), which are derived from observation Eq. (5-2) respectively. Then, a and b are calculated from the simultaneous linear equations Eqs. (5-14) and (5-15).

In the case of $y=ax$ (Fig. 5-2), the best estimated value a_{best} and its error σ_a of Eq. (5-16) are obtained from Eqs. (5-17) and (5-18) respectively. The σ_y defined by Eq. (5-18) is calculated from Eq. (5-19). Here, in Eq. (5-19), $(N-2)$ of Eq. (5-13) becomes $(N-1)$, as the number of unknown parameters is 2 at $y=ax+b$ and that of

式(5-18)の σ_y は式(5-19)で求めることができる．$y=ax+b$では未知定数が2個であったが，$y=ax$ では未知定数が1個なので，式(5-19)では$(N-1)$となる．

次に，計測結果を多項式で近似するときの次数の決定方法について述べる．一例として，川の流れを計測したとき，水深と流速がTable 5-1の結果を得た場合について考える．各計測点ごとの一次差分δ^1，二次差分δ^2を順次求めるとTable 5-1のようになり，これを階差表と呼ぶ．計測点（Table 5-1の水深）が等間隔でない場合は，計測間隔を考慮する必要がある．Table 5-1では，二次差分δ^2がほぼ一定となり，三

Table 5-1　Determination of degree (Example of flow velocity distribution)

Distance from surface m	Flow velocity m/s	δ^1	δ^2	δ^3	δ^4
0	1.9720				
		−0.0350			
0.1	2.0070		−0.0119		
		−0.0231		0.0042	
0.2	2.0301		−0.0161		−0.0052
		−0.0070		0.0094	
0.3	2.0371		−0.0255		0.0300
		0.0185		−0.0206	
0.4	2.0186		−0.0049		−0.0288
		0.0234		0.0082	
0.5	1.9952		−0.0131		0.0037
		0.0365		0.0045	
0.6	1.9587		−0.0176		0.0090
		0.0541		−0.0045	
0.7	1.9046		−0.0131		−0.0085
		0.0672		0.0040	
0.8	1.8374		−0.0171		
		0.0843			
0.9	1.7531				
Root mean square value		0.0454	0.0159	0.0097	0.0179

$y=ax$ is 1.

Next is a definition of the degree method to reveal experimental results by polynomial expression. As a reference, the procedure for using the result of flow velocity distribution, which is the relation between the distance from surface and flow velocity, as shown in Table 5-1, follows. The first order difference δ^1 and second order difference δ^2 of measured values are shown in Table 5-1, which is called the finite difference table. If the steps of measuring points are not constant, the step width should be considered. In Table 5-1, the second order differences δ^2 are nearly constant and the sum of the squares of third order differences has a minimum, so a

次差分 δ^3 の二乗平均平方根が最小となっていることから二次式で近似するのが妥当であるといえる．

計測した 2 つの物理量 (x, y) の N 組の計測値 (x_i, y_i) $(i=1, 2, 3, \ldots, N)$ の相関関係の有無を調べるのに，式(5-20)で定義される線形相関係数 r が用いられる．また，式(5-21)で定義される σ_{xy} は共分散と呼ばれる．$\bar{x}, \bar{y}$ は式(5-22)，(5-23)で定義される．線形相関係数は $-1 \leqq r \leqq 1$ の値をとり，絶対値が大きいほど線形相関が強く，絶対値が小さいほど線形相関が弱いことを意味する．また，$r \geqq 0$ の場合を正の相関，$r \leqq 0$ の場合を負の相関と呼ぶ．ただし，線形相関係数を用いて定量的に相関を議論する場合には計測値の個数（回数）に注意を要する．例えば 2 回のみの計測では $r=1$ となるが，端的に両者に相関関係があると結論付けるのは尚早である．一方，相関がない (x, y) の場合には，無限回計測した場合には $r=0$ となるが，少数回の計測では $r=0$ となることはほとんどあり得ない．ここで，相関がない (x, y) を仮定して考えると，相関がない (x, y) の場合でも r がある値 r_0 よりも大となる確率 $P_N(|r| \geqq r_0)$ を式(5-24)により計算することができ，それを Table 5-2 に示す．Table 5-2 より，5 回計測した場合に $r=0.8$ であっても，無相関である確率が 10%存在することを意味す

$$r = \frac{\sum_{i=1}^{N}(x_i - \bar{x})(y_i - \bar{y})}{\sqrt{\sum_{i=1}^{N}(x_i - \bar{x})^2 \sum_{i=1}^{N}(y_i - \bar{y})^2}} \quad (5\text{-}20)$$

$$\bar{x} = \frac{1}{N}\sum_{i=1}^{N} x_i \quad (5\text{-}22)$$

$$\sigma_{xy} = \frac{1}{N}\sum_{i=1}^{N}(x_i - \bar{x})(y_i - \bar{y}) \quad (5\text{-}21)$$

$$\bar{y} = \frac{1}{N}\sum_{i=1}^{N} y_i \quad (5\text{-}23)$$

quadratic equation should be reasonable to describe the results of Table 5-1.

In order to investigate the relationship of two values (x, y) from N sets of measured values (x_i, y_i) $(i=1, 2, 3, \ldots, N)$, a linear correlation coefficient r defined by Eq. (5-20) is normally used. The σ_{xy} defined by Eq. (5-21) is called the covariance. $\bar{x}$ and $\bar{y}$ are defined by Eqs. (5-22) and (5-23). The r has a value of $-1 \leqq r \leqq 1$, with a larger absolute value revealing strong linear correlations and a smaller absolute value meaning weak linear correlations. Positive correlations exist at $r \geqq 0$ and negative correlations at $r \leqq 0$. It must be noted that the number of measurements is very important in assessing the correlation quantitatively by the linear correlation coefficient r. It is clear that $r=1$ from two sets of measured values, however, it should be considered very carefully. If there is no correlation between two values with many measurements, r equals 0, however, r is not 0 with a limited number of measurements. The probability of no correlation $P_N(|r| \geqq r_0)$ is obtained by Eq. (5-24) and is shown in Table 5-2. Table 5-2 shows that probability of no correlation of r

る．$P_N(|r|\geqq r_0)$ が小さいほど，両者が無相関である確率は小さい．一般に $P_N(|r|\geqq r_0)\leqq 5\%$ の場合には両者の相関は有意であるということができ，$P_N(|r|\geqq r_0)\leqq 1\%$ の場合には高度に有意であるということができる．したがって，5回計測して $r=0.8$ の場合には $P_N(|r|\geqq r_0)=10\%\geqq 5\%$ なので有意であるとはいえないが，15回計測して $r=0.6$ の場合には上記の線形相関係数よりも小さいが $P_N(|r|\geqq r_0)=1.8\%\leqq 5\%$ なので有意であるといえる．

Table 5-2　Probability of decorrelation $P_N(|r|\geqq r_0)$〔%〕

N	r_0												
	0	0.1	0.2	0.3	0.4	0.5	0.6	0.7	0.8	0.85	0.9	0.95	1
3	98	92	85	79	72	65	57	49	39	33	27	18	0
4	100	90	80	70	60	50	40	30	20	15	10	5.0	0
5	100	87	75	62	51	39	29	19	10	6.8	3.8	1.4	0
6	100	85	70	56	43	31	21	12	5.6	3.2	1.5	0.4	0
7	100	83	67	51	37	25	15	8.0	3.1	1.6	0.6	0.1	0
8	100	81	64	47	33	21	12	5.3	1.7	0.8	0.2		0
9	100	80	61	43	29	17	8.8	3.6	1.0	0.4	0.1		0
10	100	78	58	40	25	14	6.7	2.4	0.6	0.2			0
11	100	77	56	37	22	12	5.1	1.7	0.3	0.1			0
12	100	76	53	34	20	9.8	3.9	1.1	0.2				0
13	100	75	51	32	18	8.2	3.0	0.8	0.1				0
14	100	74	49	30	16	6.9	2.3	0.5	0.1				0
15	100	72	48	28	14	5.8	1.8	0.4					0
16	100	71	46	26	13	4.9	1.4	0.3					0
17	100	70	44	24	11	4.1	1.1	0.2					0
18	100	69	43	23	10	3.5	0.9	0.1					0
19	100	69	41	21	9.0	2.9	0.7	0.1					0
20	100	68	40	20	8.1	2.5	0.5	0.1					0

$$P_N(|r|\geqq|r_0|)=\frac{2\Gamma((N-1)/2)}{\sqrt{\pi}\,\Gamma((N-2)/2)}\int_{|r_0|}^{1}(1-r^2)^{(N-4)/2}dr \qquad (5\text{-}24)$$

$=0.8$ at $N=5$ is 10%. The smaller $P_N(|r|\geqq r_0)$ means a lower probability of no correlation between two parameters. It can be said that the relationship is significant at $P_N(|r|\geqq r_0)\leqq 5\%$, and highly significant at $P_N(|r|\geqq r_0)\leqq 1\%$. For example, $r=0.8$ at $N=5$ is not significant, but $r=0.6$ at $N=15$ is highly significant, as $P_N(|r|\geqq r_0)=10\%\geqq 5\%$ of $r=0.8$ at $N=5$ and $P_N(|r|\geqq r_0)=1.8\%\leqq 5\%$ of $r=0.6$ at $N=15$.

● 5.2 時系列データの処理

時系列データ $x(t)$ を時刻 t_1 から時刻 t_2 まで計測した場合，時間平均 $\overline{x}$ は式(5-25)で定義される．これを直流成分（DC 成分）と呼ぶこともある．$x(t)$ に正の値と負の値が含まれる場合には式(5-26)で定義される二乗平均平方根 x_{RMS} を用いて評価する場合が多い．

次に，時系列データ $x(t)$ の周期性の評価について考える．簡単にするために $\overline{x}=0$ の場合について考える．時刻 $-T/2$ から時刻 $T/2$ まで計測するとし，時間 T が十分長い場合には，時間パラメータ τ を考えると，時系列データ $x(t)$ と $x(t+\tau)$ の自己相関関数 $C(\tau)$ は，式(5-21)より式(5-27)で定義される．なお，$x(t)$ は，個々の振動数 ω_n を有する関数の和として係数 a_n，b_n を用いて式(5-28)～(5-30)のようにフーリエ級数の形で表記することが可能である．これらは第 4 章のフーリエ級数を記述した式において $\omega=2\pi n\dfrac{t}{T}$ と書き換えた式と同じである．関数 $x(t)$ が滑らかな連続関数で

$$\overline{x}=\frac{\int_{t_1}^{t_2}x(t)\,dt}{\int_{t_1}^{t_2}dt} \quad (5\text{-}25)$$

$$x_{RMS}=\sqrt{\frac{1}{t_2-t_1}\int_{t_1}^{t_2}x(t)^2dt} \quad (5\text{-}26)$$

$$C(\tau)=\lim_{T\to\infty}\frac{1}{T}\int_{-\frac{T}{2}}^{\frac{T}{2}}x(t)\,x(t+\tau)\,dt \quad (5\text{-}27)$$

$$x(t)=\frac{a_0}{2}+\sum_{n=1}^{\infty}(a_n\cos\omega_n t+b_n\sin\omega_n t) \quad (5\text{-}28)$$

$$a_n=\frac{2}{T}\int_{-\frac{T}{2}}^{\frac{T}{2}}x(t)\cos(n\omega t)\,dt \quad (5\text{-}29)$$

$$b_n=\frac{2}{T}\int_{-\frac{T}{2}}^{\frac{T}{2}}x(t)\sin(n\omega t)\,dt \quad (5\text{-}30)$$

● 5.2 Processing of time-series data

The time-averaged value $\overline{x}$ of time-series data $x(t)$ from t_1 to t_2 is defined by Eq. (5-25), and called the direct current (DC) component. When $x(t)$ includes both positive and negative values, the root mean square value x_{RMS}, as defined by Eq. (5-26), is used.

Next, we turn to the periodicity of $x(t)$. Now, $x(t)$ of $\overline{x}=0$ is considered. If the time-dependent parameter t was concerned from $-T/2$ to $T/2$ at large T, the auto-correlation function $C(\tau)$ between $x(t)$ and $x(t+\tau)$ is defined by Eqs. (5-27), from (5-21). Here, $x(t)$ can be expressed by Eqs. (5-28) through (5-30), with each having the oscillation frequency ω_n. This is similar to the equation that describes a Fourier equation in Chapter 4, with $\omega=2\pi n\dfrac{t}{T}$. Since the Fourier equation can be expressed by Eq.

あればフーリエ積分は式(5-31)で与えられるので，オイラーの表記法を用いれば式(5-34)と書くことができ，$x(t)$のフーリエ変換$X(\omega)$は式(5-35)となる．ここで，パワースペクトル$S(\omega)$を式(5-36)のように定義すると，パワースペクトル$S(\omega)$は自己相関関数$C(\tau)$のフーリエ変換として求めることができる．なお，$X(\omega)$は各振動数における振幅であるが，$S(\omega)$は振動数におけるエネルギーである．以上要するに，時系列データ$x(t)$は，種々の振動数ω_nからなる時系列データの和として考えることができ，それぞれの振動数ω_nの振幅$X(\omega)$は$x(t)$のフーリエ変換によって求めることができる．

実際に時系列データからフーリエ変換により周波数解析する場合には，高速フーリエ変換が用いられる．なお，実際の計測においては，計測時間は有限なので，解析時には矩形窓，ハニング窓，フラットトップ窓などの窓関数が用いられる．窓関数には，周波数分解能が良く，ダイナミックレンジが広いことが求められるが，両立すること

$$x(t)=\frac{1}{2\pi}\int_{-\infty}^{\infty}\{A(\omega)\cos\omega t+B(\omega)\sin\omega t\}\,d\omega \tag{5-31}$$

$$A(\omega)=\int_{-\infty}^{\infty}x(t)\cos\omega t\,dt \tag{5-32}$$

$$B(\omega)=\int_{-\infty}^{\infty}x(t)\sin\omega t\,dt \tag{5-33}$$

$$x(t)=\int_{-\infty}^{\infty}X(\omega)\,e^{j\omega t}\,d\omega \tag{5-34}$$

$$X(\omega)=\frac{1}{2\pi}\int_{-\infty}^{\infty}x(t)\,e^{-j\omega t}\,dt \tag{5-35}$$

$$S(\omega)=\lim_{T\to\infty}\frac{2\pi\,|X(\omega)|^2}{T} \tag{5-36}$$

$$S(\omega)=\frac{1}{2\pi}\int_{-\infty}^{\infty}C(\tau)\,e^{-j\omega t}\,d\tau \tag{5-37}$$

(5-31), when $x(t)$ is a smooth and continuous function, Eq. (5-34) provides the Euler expression, and the Fourier transformation $X(\omega)$ is revealed by Eq. (5-35). When the power spectrum $S(\omega)$ is defined by Eq. (5-36), $S(\omega)$ can be obtained from $C(\tau)$ by Fourier transformation. Here, $X(\omega)$ means the amplitude at each frequency and $S(\omega)$ shows the energy at each frequency. Namely, $x(t)$ can be considered the sum of time-series data having the frequency of ω_n, and $X(\omega)$ can be obtained by Fourier transformation of $x(t)$.

Fast Fourier transformation (FFT) is used at the experimental level. As the measuring length is limited, window functions like the rectangular window, hamming window, and flat-top window are used. Although window functions require

は困難で，窓関数により周波数分解能と周波数領域に優劣が存在するので，計測時に選択する場合は注意を要する．なお，解析周波数領域は，4.3 節のサンプリング定理で述べたように，下限周波数はサンプリング長に，上限周波数はサンプリング間隔に依存する．

時系列データにおいてノイズが存在する場合には，14.3 節の信号をきれいにする回路で述べるようにアナログ回路によりノイズを除去することも可能であるが，デジタル化した時系列データ $x(i)$ の場合，移動平均化処理や積算平均化処理によりノイズを除去することも可能である．式(5-38)には，n 点の $x(i)$ から，$x(i)$ の前後の $2k+1$ 点を移動平均して求める $g(i)$ を示す．$i=1\sim k$ と $i=n-k+1\sim n$ では i の前後に $2k+1$ 点を確保できないため $g(i)$ を算出できないので考慮する必要がある．また，式(5-39)には，n 点を有する $x(i)$ を K 回計測した場合に積算平均化処理して求める $h(i)$ を示す．

時系列データにおいてデータを整理して実験式を求める際に，t の増大による Δx

$$g(i)=\frac{1}{2k+1}\sum_{j=-k}^{k}x(i+j)\quad (i=k+1, k+2, \ldots, n-k) \tag{5-38}$$

$$h(i)=\frac{1}{K}\sum_{j=1}^{K}x_j(i)\quad (i=1, 2, \ldots, n) \tag{5-39}$$

・Increment of Δx is constant with increase of t

$$x(t)=a\,t+b \tag{5-40}$$

・Increment of Δx is proportional to t with increase of t

$$x(t)=at^2+bt+c \tag{5-41}$$

・Increment of Δx is proportional to x with increase of t

$$x(t)=ae^{bt} \tag{5-42}$$

・Increment of Δx is proportional to $(1-x/a)$ with increase of t

$$x(t)=a(1-e^{-bt}) \tag{5-43}$$

characteristics of good frequency resolution and good dynamic range, it is very difficult to have both characteristics, in that case, the required characteristics should be studied. As shown in Chapter 4, lower frequencies depend on the sampling length and higher frequencies depend on the sampling step.

When the time series data has noise, it is possible to clear it by analog circuit, as shown in Section 14.3. It is also possible to remove the noise by processing digital $x(i)$ through a moving average and/or a cumulative average. The moving average $g(i)$ from n points before and after $x(i)$ is shown in Eq. (5-38). Eq. (5-39) reveals the cumulative average $h(i)$ from the number of measurements K.

の増分を考慮すると，それぞれ式(5-40)～(5-43)のような形になる．

● 5.3 有効数字と誤差の伝播

計測において意味のある数字を有効数字という．意味のない数値の羅列は，数値や桁を間違える原因になるので避けるべきである．通常，誤差は有効数字以下の値を四捨五入して有効数字1桁に丸め，最良推定値は誤差の有効数字の最終桁と同じ位置になるように四捨五入して表す．誤差の有効数字を1桁にするのは，計測値の標準偏差σを誤差として$\pm\sigma$と表記する場合が多いが，計測値が正規分布に従う場合でも$\pm\sigma$の範囲に真の値が存在する確率はたかだか68%である．よって誤差の2桁目にあまり意味がない．この例外としては，四捨五入した有効数字が「1」になる場合である．有効数字が四捨五入により1となるのは0.5～1.4の場合であり，0.5と1.4では約3倍の差がある．したがって，有効数字が1の場合は，その1つ下の桁まで表記すべきである．

計測結果を用いて加減乗除の計算を行う場合には，誤差の伝播を考慮する必要があ

$$p=x+\cdots+z-u-\cdots-w \tag{5-44}$$

$$\delta p=\sqrt{(\delta x)^2+\cdots+(\delta z)^2+(\delta u)^2+\cdots+(\delta w)^2} \tag{5-45}$$

$$q=\frac{x\times\cdots\times z}{u\times\cdots\times w} \tag{5-46}$$

$$\frac{\delta q}{|q|}=\sqrt{\left(\frac{\delta x}{x}\right)^2+\cdots+\left(\frac{\delta z}{z}\right)^2+\left(\frac{\delta u}{u}\right)^2+\cdots+\left(\frac{\delta w}{w}\right)^2} \tag{5-47}$$

In order to establish experimental formulas, Eqs. (5-40) through (5-43) are proposed, considering the increment of Δx with increase in t.

● 5.3 Significant digits and propagation of error

In measurement, a number having a meaning is a significant digit. Enumeration of numbers might cause a misunderstanding of value and the cancellation of significant digits. Normally, the error is rounded off to the last digit and the best estimated value are shown as the same digit of the error. This is why the error is shown in one digit; the probability of true value exists within standard deviation $\pm\sigma$ is 68%, even though the measured value follows the normal distribution. Thus, the secondary digit is not required. However, when the significant digit is "1", that is exceptional and a second digit is required. When "1" is obtained by rounding off, it can range from 0.5 to 1.4, and 1.4 is nearly three times 0.5. Thus, when the significant digit is "1", the secondary digit is required.

る．得られた測定値$x, ..., z, u, ..., w$のそれぞれの誤差が$\delta x, ..., \delta z, \delta u, ..., \delta w$であり，これらの誤差が互いに独立でランダムであるとき，式(5-44)のように和と差の計算によりpを求める場合，その誤差δpは式(5-45)に示すように個々の絶対誤差の二乗和で求めることができる．また，得られた測定値$x, ..., z, u, ..., w$から式(5-46)のような積と商の計算によりqを求める場合，その誤差δqは式(5-47)に示すように個々の相対誤差の二乗和で求めることができる．

【演習問題】

5-1）ばねばかりのばね定数を求めるために，重りをつけてばねの長さを計測した結果，重りの質量とばねの長さはTable 5-3のとおりだった．ばね定数とその誤差を求めよ．

5-2）以下の測定結果を適当な有効数字を用いて，適切な形式で明確に表現せよ．

(1) 0.01256±0.0056789 m，(2) 21.049±1.45679 s,

Table 5-3　Relation between weight of mass and length of spring

Weight of mass kg 重りの質量 kg	1	2	3	4
Length of spring cm ばねの長さ cm	6	8	9	11

In calculating addition, subtraction, multiplication, and division, propagation of error should be considered. When the measured values are $x, ..., z, u, ..., w$ and their errors are $\delta x, ..., \delta z, \delta u, ..., \delta w$, and the errors are independent events and random, the error δp of p calculated by Eq. (5-44), such as addition and subtraction, is defined by the absolute error, as shown in Eq. (5-45). On the other hand, the error δp of p calculated by Eq. (5-46), such as multiplication and division, is defined by relative error, as shown in Eq. (5-47).

【Problems】

5-1) In order to measure the spring constant of a spring, the spring length was measured, changing with weight of mass. With the results as shown in Table 5-3, obtain the spring constant and its error.

5-2) Describe suitable formations of the following measured values, considering significant digits.

(1) 0.01256±0.0056789 m，(2) 21.049±1.45679 s,

(3) $-7.642\times10^{8}\pm2.533\times10^{7}$ Pa

5-3) 60人のクラスで，レポートの評価点と期末試験の成績点の相関係数が0.4であった．レポートの評価点と期末試験の成績点に相関があるといえるか．その理由を述べよ．

(3) $-7.642\times10^{8}\pm2.533\times10^{7}$ Pa

5-3) The correlation coefficient between points on reports and the score on the final exam was 0.4. The number of students in the class was 60. Is there any correlation between the points on reports and the score on the final exam? Explain your reasoning.

第6章　変位と変形の計測

物体に力を加えると，その位置が変わったり，形状が変化したりする．物体の位置の変化を変位，また形状の変化を変形と呼ぶ．変位と変形はSI単位系の基本単位「長さ」に直結するものである．また，ニュートンの運動方程式を中心とする剛体の力学では変位が，弾性力学や塑性力学では変形がそれぞれ基本的な物理量となっているように，変位と変形はロボットなどのシステムにとって重要なパラメータである．なお，加速度や力などの計測は変位と変形を介して行われる場合が多く，変位と変形の計測はこれらの量の計測の基本になっているといえる．

● 6.1　変位の計測

計測システムにおける信号変化の観点から，変位計測法を電磁気的原理に基づくものと光学的原理に基づくものに分けることができる．Fig. 6-1にはそれぞれの原理に基づく主な計測機の名称およびそれぞれの計測機で変位が変換される量の名称を示し

Chapter 6　Measurement of Displacement and Deformation

Changes in position and shape are referred to as displacement and deformation respectively. Displacement and deformation are directly linked to the base unit of length in the SI system. Displacement is a fundamental quantity in rigid body mechanics based on Newton's laws of motion, while deformation is a fundamental quantity in elastic and plastic mechanics. Both are therefore important parameters for systems such as robots. They also form the foundation for the measurement of related quantities such as acceleration and force.

● 6.1　Measurement of displacement

The methods of measuring displacement can be classified into electromagnetic and optical methods, as shown in Fig. 6-1. Displacement is converted into electrical quantities such as electrical resistance or current by electromagnetic methods.

ている．電磁気式変位計測機の場合は，変位などを直接電気抵抗や電流などの電気量に変換している．一方，抵抗や電流などの電気量はそのまま利用しにくいため，後続の電気回路を使ってさらに電圧に二次変換するようにしている．このような変換回路を含めた計測回路は第14章で述べる．また，光学式変位計測機の場合は，一旦変位を光学的な信号に変換してから第12章に記載されるフォトダイオードなどの光電変換デバイスを使って光学的信号から電気信号に変換することになっている．

以下では電磁気式変位計測機の代表例として静電容量型変位計，また光学式変位計測機の代表例として三角測量法レーザ変位計を取り上げ，その基本原理と主な性能について述べる．

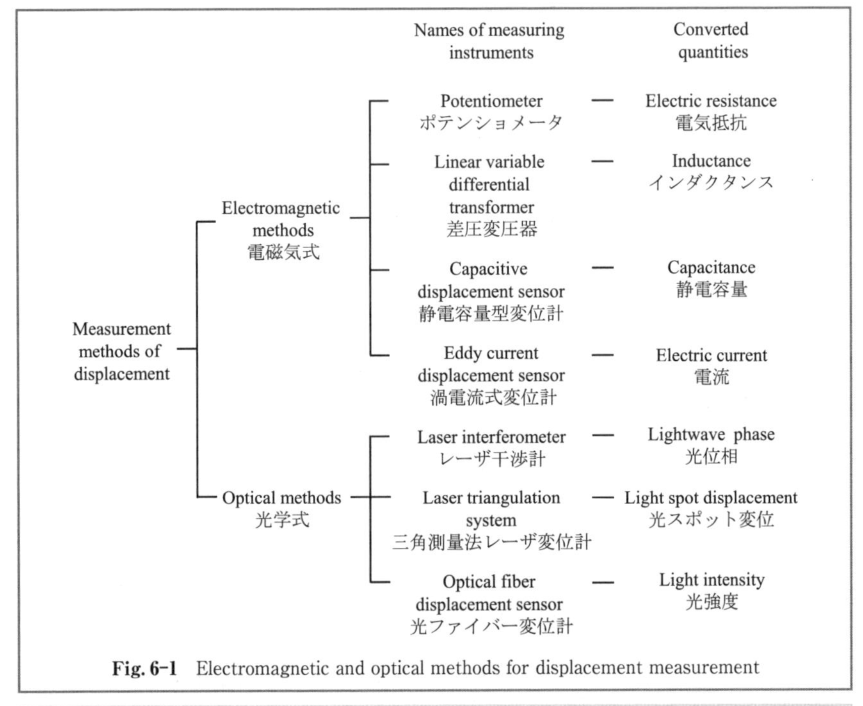

Fig. 6-1 Electromagnetic and optical methods for displacement measurement

Quantities other than voltage are then converted into voltage by using electrical circuits, as described in Chapter 14, for easier usage. In optical methods, displacement is converted into electrical signals by using optoelectronic devices such as the photodiodes presented in Chapter 12.

In what follows, the capacitive displacement sensor and the laser triangulation displacement sensor are presented as representatives of the two methods. The basic principles and main characteristics of each are described.

静電容量型変位計　静電容量型変位計はコンデンサの原理に基づくものである．Fig. 6-2 に 2 枚の金属電極板からなるセンサの模式図を示す．金属電極板が xy 面に平行に配置されているとする．電極間の静電容量 C（単位ファラド F）は式(6-1)で表すことができる．ここで，Q は電極に蓄えられている電荷（単位クーロン C），V は電極間電位差（単位ボルト V）である．

電極間に大気など誘電率 ε の絶縁物が均一に充填されているとすると，ガウスの法則より電極間の電界 E（単位 V/m），電圧 V，静電容量 C はそれぞれ式(6-2)，(6-3)，(6-4)のように求められる．ここで，S_{xy} は金属電極が向き合っている部分の面積（単位 m^2）であり，d_z は電極間間隔（単位 m）である．式(6-4)から分かるように，電極間の静電容量が電極の面積に比例し，電極間距離に反比例する．以下ではこれを利用して，電極間の変位を求める．

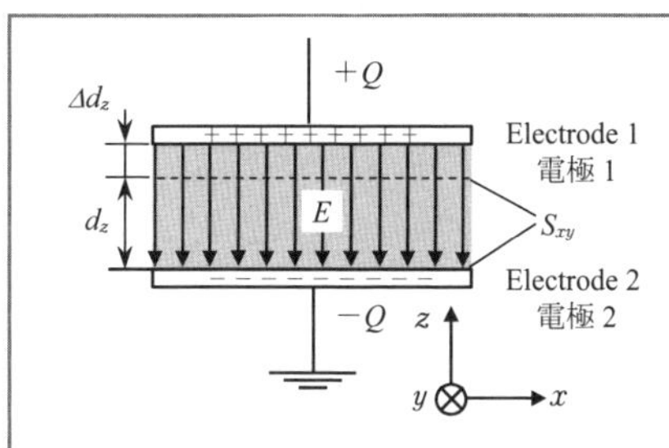

Fig. 6-2　Conversion of z-directional vertical displacement to capacitance change in a capacitor

$$C=\frac{Q}{V} \tag{6-1}$$

$$E=\frac{Q}{\varepsilon S_{xy}} \tag{6-2}$$

$$V=Ed_z=\frac{Q}{\varepsilon S_{xy}}\,d_z \tag{6-3}$$

$$C=\frac{\varepsilon S_{xy}}{d_z} \tag{6-4}$$

Capacitive displacement sensor　A capacitive displacement sensor is based on a capacitor. Fig. 6-2 shows a schematic of the sensor composed of two metal electrode plates. The electrode plates are placed parallel to the xy plane. The capacitance C between the two plates with a unit of farad (F) can be expressed by Eq. (6-1). Here, Q is the charge accumulated on the electrode with a unit of coulomb (C) and V is the potential difference between the electrodes with a unit of voltage (V).

Assume that a dielectric insulating material with a dielectric constant ε is inserted between the plates. The electric field E with unit V/m, electric voltage V, and capacitance C can be obtained in Eqs. (6-2), (6-3), and (6-4) respectively. Here, S_{xy} is the overlapping area of the plate with a unit of m^2 and d_z is the gap between the plates with a unit of m. As Eq. (6-4) shows, C is proportional to S_{xy} and inversely proportional to d_z. This is utilized to measure the displacement between the electrodes, as detailed in what follows.

When a displacement Δd_z is applied to Electrode 1 with respect to Electrode 2, as

Fig. 6-2には電極2が電極1に対してz方向にΔd_zだけ変位した場合も示されている．変位Δd_zに伴って，静電容量Cが式(6-5)に示すC_1に変化する．CとC_1の差をΔCとすると，ΔCは式(6-6)のように表せる．変位Δd_zが初期電極間間隔d_zより十分小さいとすると，式(6-6)が式(6-7)のように書き直せる．つまり，変位Δd_zに伴う電極間間隔変化の割合が静電容量変化の割合にほぼ等しいことが分かる．これを利用すれば，ΔCからΔd_zを求めることができる．

また，Fig. 6-3には電極2が電極1に対してx方向にΔd_xだけ変位した場合を示す．ここで電極板が長方形の形をしており，2辺の長さがそれぞれd_x，d_yであるとする．変位Δd_xに伴い，静電容量がCからをC_2に変わるとすると，CとC_2をそれぞれ式(6-8)と(6-9)に表すことができる．CとC_2の差をΔC_2とすると，ΔC_2は式(6-

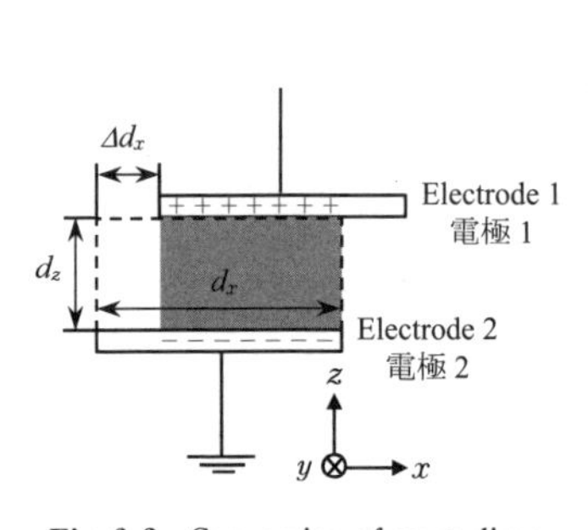

Fig. 6-3 Conversion of an x-directional horizontal displacement to capacitance change in a capacitor

$$C_1=\frac{\varepsilon S_{xy}}{d_z+\Delta d_z}=C+\Delta C \tag{6-5}$$

$$\frac{C}{C+\Delta C}=\left(1+\frac{\Delta d_z}{d_z}\right) \tag{6-6}$$

$$\frac{\Delta C}{C}\approx\frac{\Delta d_z}{d_z} \tag{6-7}$$

$$C=\frac{\varepsilon d_x d_y}{d_z} \tag{6-8}$$

$$C_2=\frac{\varepsilon(d_x-\Delta d_x)d_y}{d_z} \tag{6-9}$$

$$\Delta C_2=C\left(1-\frac{d_x-\Delta d_x}{d_x}\right) \tag{6-10}$$

$$\frac{\Delta C_2}{C}=\frac{\Delta d_x}{d_x} \tag{6-11}$$

shown in Fig. 6-2, the capacitance between the two plates will change from C to C_1. Assuming the difference between C and C_1 is ΔC, ΔC can be written in Eq. (6-6), and then rewritten in Eq. (6-7) if Δd_z is small enough when compared with the initial electrode gap d_z. It can be seen that the ratio of change in the capacitance is approximately equal to that in the electrode gap, based on which the displacement Δd_z can be evaluated.

Fig. 6-3 shows the case when an x-directional horizontal displacement Δd_x is applied to Electrode 1 with respect to Electrode 2. The electrodes are assumed to have a rectangular shape with a length d_x and a width d_y. The capacitance between the two plates will change from C to C_2, as expressed in Eqs. (6-8) and (6-9),

10)のような形に表せる．式(6-11)より，変位 Δd_x と電極長さ d_x の比が静電容量変化の割合に等しいことが分かる．これを利用して，z 軸方向と同様に静電容量の変化から変位 Δd_x を求めることができる．

Fig. 6-4 に示すように，実際の静電容量型変位計では，コンデンサの 1 つの電極をセンサ側，もう 1 つの電極はターゲット側にして，センサとターゲット間の相対変位を計測している．なお，センサには検出部のほか，ガードリング部も設けられるのが一般的になっており，両者は絶縁されている．センサ電極部とターゲット間の電気力線が漏れないようにガードリング部でセンサ電極部を囲む構造となっている．それによって，センサ電極部では電気力線が平行となり，センサ出力の高い直線性が得られる．

市販静電容量型変位計の特性は Table 2-2 に示されている．なお，市販センサの形は円筒状のものが多く，サイズは長さが 10 mm～100 mm，直径 1 mm～10 mm 程度，

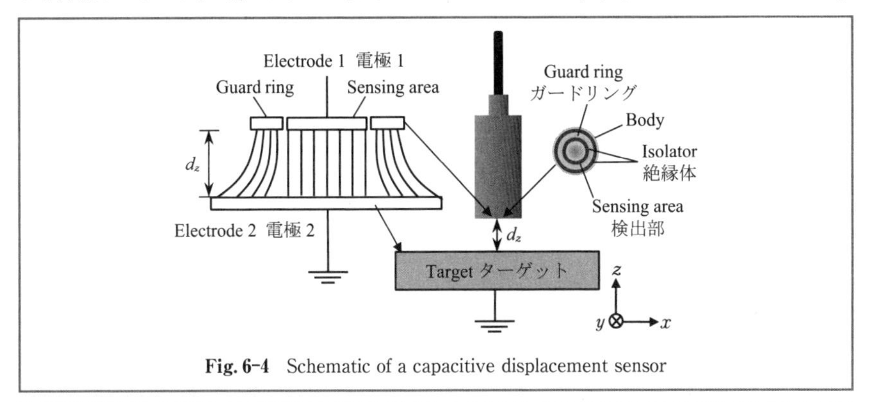

Fig. 6-4　Schematic of a capacitive displacement sensor

respectively. The difference ΔC_2 between C and C_2 is thus written in Eq. (6-10). As can be seen in Eq. (6-11), the ratio of change in the capacitance is equal to the ratio of the displacement Δd_x to the electrode length d_x. Similarly, Δd_x can be obtained from the change of the capacitance.

Fig. 6-4 shows a schematic of a practical capacitive displacement sensor that measures the z-directional displacement between the sensor and a target. Electrode 1 is the sensing electrode on the sensor head and Electrode 2 is the target. A guard ring is added to surround the sensing electrode. The guard ring and the sensing electrode are isolated with an insulating material. This design shields the sensing electrode from boundary effects so that the electric lines of force can be kept straight for a good linearity of sensor output.

The characteristics of a commercial capacitance displacement sensor are shown in Table 2-2 of Chapter 2. This sensor type often has a cylindrical probe design. The

測定範囲は小さいものでは10 μm程度，大きいものでは1 mm程度となっている．静電容量型センサは非接触，高精度，高速，小型，取り扱いが容易などの特徴があり，軸受やモータなどの動特性評価，精密ステージの位置決めと姿勢制御，加工面の形状計測などに広く用いられている．測定範囲を10 μm以内に限定すれば，サブnmの高い分解能も実現可能なので，ナノ計測の分野でもよく利用される．基本的に金属と半導体のような導電体のターゲットにしか対応できないが，ガラスなどの絶縁体が使えるように工夫されているものもある．

一方，シリコン材料をベースとするマイクロマシン加工技術によって，微小サイズの静電容量型センサが製作できるため，加速度センサや静電アクチュエータなど変位検出が必要なデバイスに組み込みセンサとして利用されている．

三角測量法レーザ変位計　三角測量法は三角形の幾何学的関係に基づき，三角形の内角を利用して辺の長さを求める方法であり，古くから長さや距離の測量に利用されている．Fig. 6-5に示すように，三角測量法では，直角三角形AOBにおいて，辺ABの長さhは辺AOの長さpとAO，BOのなす角αから式(6-12)のように求めること

length of the probe ranges from 10 mm to 100 mm, while its diameter ranges from 1 mm to 10 mm. The measurement range of displacement is typically larger than 10 μm, and some can reach up to 1 mm. Capacitive displacement sensors have the advantages of avoiding contact, high accuracy, fast speed, compact size, and ease of use, and are widely used in the dynamic measurement of bearings and motors, the positioning of precision stages, the profile measurement of machined surfaces, etc. Since sub-nanometric resolution of the displacement can be achieved with a limited range of 10 μm, capacitive displacement sensors are also often used in nanometrology. Basically, only metal targets can be measured, though some commercial sensors can also measure isolators such as glass.

On the other hand, capacitive displacement sensors with micro sizes can be made using silicon machining processes. Such micro sensors are embedded in acceleration sensors and electrostatic actuators for displacement measurement.

Laser triangulation sensor　In the triangulation method, the length of a side of a triangle is measured by utilizing the internal angle of the triangle based on the geometrical properties of triangles. It has long been employed for measurement of

ができる．あるいは，ある物体が点Aから点Bに移動したとき，その変位量Δhは同様に式(6-13)より求められる．式(6-13)にある$\Delta\alpha$は∠AOBの変化量であり，角変位量とも呼ばれる．長さや距離hを測定するための伝統的な三角測量では，セオドライトなどの角度計測機を使って角度αを計測し，式(6-12)に基づいてhの計算を行っている．この方法は，hが大きい場合にも対応できる特徴がある．またレーザトラッカーなど，動的な角変位$\Delta\alpha$が計測できる角度計測機を用いれば，物体の変位Δhの計測も可能である．一方，測定範囲が限定されるが，変位Δhを測定するレーザ変位計では，角変位$\Delta\alpha$を直接計測しない形で，以下で述べる結像光学系を利用して三角測量の原理をコンパクトな形で実現している．

Fig. 6-6に示す結像光学系では，物体空間にある点AとBは結像レンズによってそれぞれ像空間にある点A′とB′に結像される．∠AOBと∠A′OB′が等しいことを利用して，三角形A′OB′において$\tan\Delta\alpha$が式(6-14)のように求められるので，それ

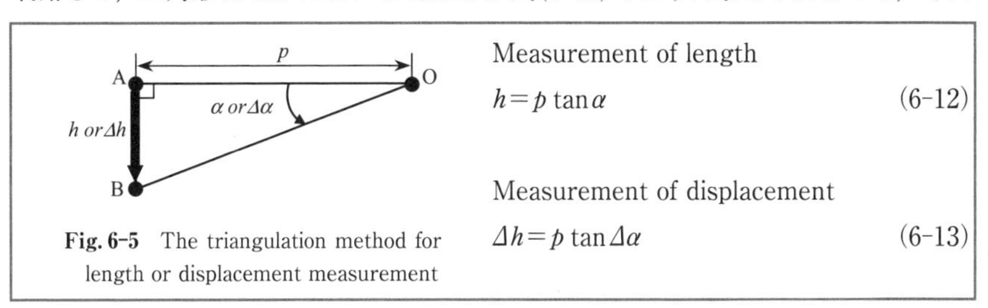

Fig. 6-5　The triangulation method for length or displacement measurement

Measurement of length

$$h = p \tan\alpha \qquad (6\text{-}12)$$

Measurement of displacement

$$\Delta h = p \tan\Delta\alpha \qquad (6\text{-}13)$$

length and distance. As shown in Fig. 6-5, the length h of side AB can be measured from the length p of side AO and the angle a between AO and BO from Eq. (6-12). Similarly, the displacement Δh of an object moving from point A to point B can be evaluated by Eq. (6-13). The $\Delta\alpha$ in Eq. (6-13) is the change of ∠AOB and is called the angular displacement. In a traditional triangulation system for measurement of length or distance h, the angle α is measured by an angle-measuring instrument such as a theodolite for the calculation in Eq. (6-12). The displacement Δh can also be measured by using an angle measuring instrument, such as a laser tracker, that can measure dynamic angular displacement $\Delta\alpha$. On the other hand, although the measurement range is limited, the displacement can be obtained without directly measuring $\Delta\alpha$ by using a laser triangulation sensor in which the principle of triangulation is realized by the imaging system in a compact structure.

In the imaging system shown in Fig. 6-6, the images of points A and B in the object space are formed at points A′ and B′ by the imaging lens. Since ∠AOB is equal to ∠A′OB′, $\tan\Delta\alpha$ in triangle A′OB′ can be evaluated by Eq. (6-14) and the

を式(6-13)に代入すれば式(6-15)のように，物体の変位 Δh を結像点の変位 $\Delta h'$ から計算できる．$\Delta h'$ は PSD や CCD 素子など光スポット変位ディテクタを用いて簡単かつ高速に計測できるため，物体変位測定のための三角測量の原理をコンパクトな形で実現し，物体変位のダイナミック計測を可能にしている．なお，式(6-16)にはレンズの中心 O から物体までの距離 p および結像点までの距離 q とレンズの焦点距離 f との関係を表している．q と p の比 k_{opt} は結像光学系の拡大倍率ともなっているので，p と q を適切に定めれば必要な拡大倍率を得ることができる．Fig. 6-7 にこの方式を利用してターゲット z 方向変位 Δd を測定するレーザ変位計の概略図を示し，式(6-17)には Δd と Δh の関係が示されている．入射角 β が 45° でない場合は，AB とレンズ光軸とのなす角が 90° からずれてしまうが，Δd が p，q に比べて十分に小さい場合は，Fig. 6-6 に示す幾何学的関係はほぼそのまま満たされる．この方式はターゲット表面からの正反射光を利用しているので，鏡面ターゲットの変位測定に適している．一方，粗面ターゲットの計測は，Fig. 6-8 に示す光学系で表面からの拡散光を利用して行う．この場合，ターゲット変位 Δd は式(6-18)のようにディテクタの出力 h'

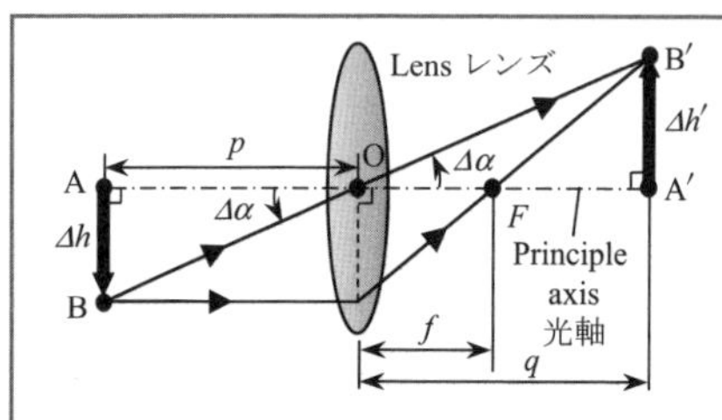

Fig. 6-6　Realization of the triangulation method by using an imaging lens

$$\tan \Delta\alpha = \frac{\Delta h'}{q} \tag{6-14}$$

$$\Delta h' = \frac{q}{p}\Delta h = k_{opt}\Delta h \tag{6-15}$$

$$\frac{1}{p} + \frac{1}{q} = \frac{1}{f} \tag{6-16}$$

displacement Δh of the object can then be calculated from the displacement $\Delta h'$ of the image in Eq. (6-15) that is obtained by substituting $\tan \Delta\alpha$ into Eq. (6-13). Since $\Delta h'$ can be easily detected by using a light position-sensing device such as a PSD or a CCD at high speed, the dynamic measurement of the object displacement Δh can be realized in a laser triangulation sensor with a compact size. Eq. (6-16) shows the relationship between p and q, which are the distances of the object and the image with respect to the lens center respectively and the focal length f of the lens. The ratio k_{opt} of q to p is the magnification of the imaging system. Fig. 6-7 shows the sensor for measuring the displacement Δd. The relationship between Δd and Δh is shown in Eq. (6-17). Although AB in Fig. 6-7 will not be perpendicular to the optical axis of the lens when β is not 45°, the geometrical relationship in Fig. 6-6 can be satisfied if Δd is small enough when compared with p and q. This type is based on specular reflection and is

と $\Delta h'$ から求めることができる．

三角測量原理のレーザ変位計は静電容量型変位計と同様に非接触式で高精度，高速，取り扱いやすいなどの特徴がある．一方，静電容量型変位計とは異なり，導電体のみならず，絶縁体の測定もできる．市販レーザ変位計の測定範囲は短いものでは 100 μm 程度，長いものでは 10 mm 程度となっている．CCD 素子や CMOS 素子など広い検出範囲を持つディテクタを用いれば，静電容量型変位計に比べて広い測定範囲が得られる．また，レンズからターゲット面の距離 d は作動距離と呼ばれるが，レーザ変位計は測定範囲に関係なく数十 mm までの長い作動距離が取れるのも大きな利点である．それに対して，静電容量型変位計の作動距離は基本的に変位測定範囲よりも短いことになっている．ただし，Fig. 6-6 から分かるように，拡大倍率 k_{opt} が大き

$$\Delta d = \Delta h \cos\beta \quad (6\text{-}17)$$

$$\Delta d = w \cdot s\left(\frac{1}{h'} - \frac{1}{h' + \Delta h'}\right) \quad (6\text{-}18)$$

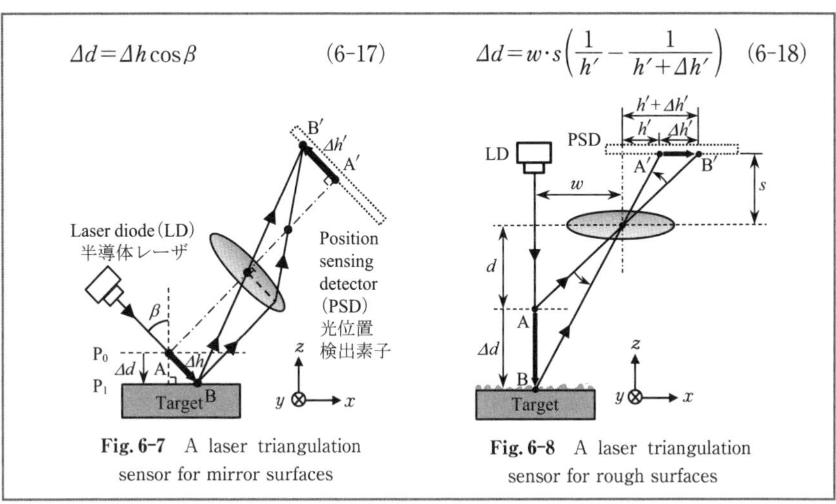

Fig. 6-7 A laser triangulation sensor for mirror surfaces

Fig. 6-8 A laser triangulation sensor for rough surfaces

suitable for measuring mirror surfaces. For rough surfaces, the sensor in Fig. 6-8 is based on diffuse reflection. In this case, the displacement Δd can be obtained from h' and Δh in Eq. (6-18).

Like the capacitive displacement sensor, the laser triangulation sensor has the advantages of avoiding contact, high accuracy, fast speed, and ease of use. Furthermore, the laser triangulation sensor can operate not only conductive surfaces but also on non-conductive surfaces. The measurement range of a commercial laser triangulation sensor is typically larger than 100 μm. Some sensors have ranges up to 10 mm, which is much higher than capacitive sensors, by using CCD or CMOS technology that offers a wide range of light position detection. The distance d from the lens to the target surface is referred to as the working distance, which can be up to

くとりにくいので，レーザ変位計の分解能は基本的にディテクタの分解能と同程度になっており，市販レーザ変位計では0.1 μm程度のものが多いのがこの原因である．

● 6.2　変形の計測

物体の変形 Δd は静電容量型変位計やレーザ変位計のような変位センサで計測することも可能であるが，センサの取り付けが困難な場合などは，ひずみゲージをターゲットに接着剤などで張り付けて計測する（Fig. 6-9）．

金属ひずみゲージは長さ l，直径 b，断面積 S の金属導線であり，その電気抵抗 R（単位 Ω）は式(6-19)で表すことができる．ρ は導線の電気抵抗率と呼ばれ，単位は Ω m である．ターゲットの変形に伴い，導線の長さと直径はそれぞれ $l-\Delta l$，$b+\Delta b$ に変化し，電気抵抗は式(6-20)のように変わる．ただし，ρ は変化しないものとする．導線の横ひずみ $\Delta b/b$ と縦ひずみ $\Delta l/l$ の比であるポアソン比 ν_s（式(6-21)）を利用して，導線の電気抵抗変化率 $\Delta R/R$ を式(6-22)と(6-23)のように導出できる．これらの式から，$\Delta R/R$ は $\Delta l/l$ に比例することが分かる．その比例係数 k_s はひず

several tens of mm regardless of the measurement range. It is much longer than the working distance of a capacitive displacement sensor. On the other hand, it is difficult to reach the large magnification k_{opt} in Fig. 6-6, and the resolution of a laser triangulation sensor is basically comparable to that of the light position-sensing detector. This is why a commercial sensor typically has a resolution of 0.1 μm.

● 6.2　Measurement of deformation

Sometimes, it is difficult to mount a displacement sensor to measure deformation. In that case, the deformation measurement can be made by gluing a strain gauge on the target, as shown in Fig. 6-9.

A strain gauge is a metal wire with length l, diameter b, and sectional area S. Its electric resistance R is expressed by Eq. (6-19). ρ is the electrical resistivity of the wire with a unit of Ω m. The length and diameter are changed to $l-\Delta l$ and $b+\Delta b$ respectively, with the deformation of the target. The change in electric resistance is written in Eq. (6-20), where ρ is assumed not to change. The Poisson's ratio of $\Delta b/b$ to $\Delta l/l$ is expressed by ν_s in Eq. (6-21), from which the rate of change $\Delta R/R$ in the electric resistance can be evaluated from Eqs. (6-22) and (6-23). It can be seen that $\Delta R/R$ is proportional to $\Delta l/l$. In Eq. (6-24), k_s is referred to as the gauge factor and is

みゲージのゲージ率と呼ばれる．式(6-24)に示されるように，ゲージ率は材料のポアソン比によって決まる．通常の金属ひずみゲージのゲージ率は２程度となっている．市販金属ひずみゲージは，薄い樹脂の電気絶縁体ベース上に格子状の金属箔をフォトエッチングで製作されたものになっている（Fig. 6-10）．Fig. 6-9 のようにひずみゲージをターゲットに接着剤で接着した場合，ひずみゲージにはターゲットと同じひずみが生じるので，ひずみゲージの出力からターゲットの変形を得ることができる（式(6-25)）．また，ひずみゲージを片持ちはりに接着し，片持ちはりの振動変位を計測することも．この場合，材料力学に基づいて，ひずみゲージの出力からはり先端の変位量を算出することができる．

金属ひずみゲージのほか，半導体の電気抵抗率が応力により変化するピエゾ抵抗効

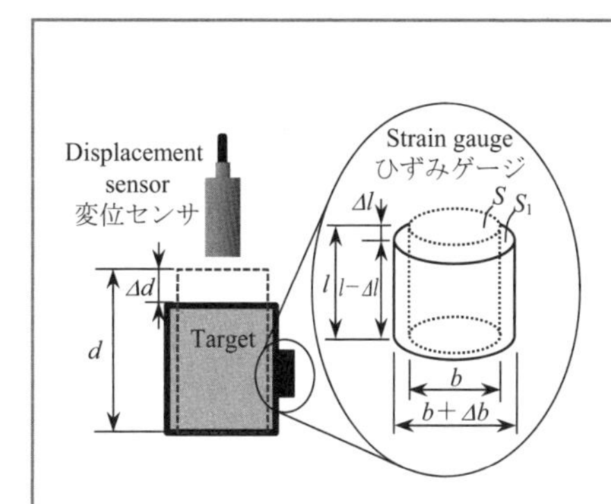

Fig. 6-9　Measurement of deformation by using a strain gauge

$$R=\rho\frac{l}{s}=\rho\frac{4l}{\pi b^2} \quad (6\text{-}19)$$

$$R-\Delta R=\rho\frac{4(l-\Delta l)}{\pi(b+\Delta b)^2} \quad (6\text{-}20)$$

$$\frac{\Delta b}{b}\approx\frac{\Delta l}{l}\nu_s \quad (6\text{-}21)$$

$$\frac{\Delta R}{R}\approx(1+2\nu_s)\frac{\Delta l}{l} \quad (6\text{-}22)$$

$$\frac{\Delta R}{R}\approx k_s\frac{\Delta l}{l} \quad (6\text{-}23)$$

$$k_s\approx(1+2\nu_s) \quad (6\text{-}24)$$

$$\frac{\Delta d}{d}=\frac{\Delta l}{l}=\frac{1}{k_s}\frac{\Delta R}{R} \quad (6\text{-}25)$$

determined by the material of the wire. The value of k_s is about 2 for a metal strain gauge. A commercial metal strain gauge is manufactured by photo-etching a metal foil onto a thin resin, as shown in Fig. 6-10. When a strain gauge is glued on a target as shown in Fig. (6-9), the target strain can be detected by a strain gauge that has the same strain as shown in Eq. (6-25). A strain gauge can also be glued on a cantilever beam to detect the vibration of the beam. The displacement of the vibrating cantilever can be calculated from the strain gauge output based on the theory of the mechanics of materials.

In addition to metal strain gauges, there are strain gauges made of semiconductor materials based on the piezoresistive effect, in which the electric resistivity changes with changes in stress. The gauge factor of a semiconductor strain gauge is higher than that of a metal strain gauge and can be used to detect small strains.

果を利用した半導体ひずみゲージもある．そのゲージ率が金属ひずみゲージよりも高いので，微小ひずみの検出に優れている．

【演習問題】

6-1）Fig. 6-2 において，大気中にある円板電極 1 の円板形検出部の直径を 2 mm とし，d_z が 100 μm から 50 μm に変わったときのセンサ静電容量の変化量 ΔC を求めよ．なお，大気の誘電率は 8.85×10^{-12} [F/m] とする．

6-2）拡大倍率 k_{opt} が 5 の Fig. 6-6 の光学系において，p と q をそれぞれ f で表せ．

6-3）Fig. 6-7 の光学系において，点 B 回りに試料面が傾斜しても反射光線が点 B′ に入る入射角 β を示せ．

6-4）銅のポアソン比 ν_s を調べ，銅線ひずみゲージのゲージ率 k_s を計算せよ．

6-5）片持ちはり先端の振動変位を計測するひずみゲージの出力感度が最大となるゲージの位置を定めよ．

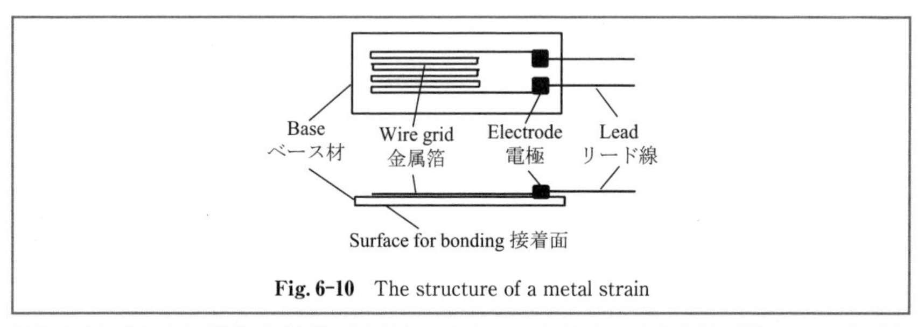

Fig. 6-10　The structure of a metal strain

【Problems】

6-1）The sensor in Fig. 6-2 has a round sensing area with a diameter of 2 mm and is used in air. Calculate the capacitance change ΔC when d_z changes from 100 μm to 50 μm. Assume that a permittivity of the air is 8.85×10^{-12} [F/m].

6-2）Using Fig. 6-6, express p and q with f when the magnification k_{opt} is 5.

6-3）In Fig. 6-7, find the angle of β for the reflected lights to go to point B′ when the target tilts about point B.

6-4）Investigate the Poisson's ratio ν_s of copper and calculate the gauge factor k_s of a copper strain gauge.

6-5）Determine the position of the strain gauge on a cantilever beam that can measure the free end displacement of the cantilever with maximum sensitivity.

第7章　速度と加速度の計測

速度と加速度は物体の運動を表す基本的な物理量である．速度は単位時間当たりの物体の位置の変化量であり，また加速度は単位時間当たりの速度の変化率であることから，第6章で述べた変位計で計測した変位の1階微分と2階微分からそれぞれ速度と加速度を求めることができる．しかし，その微分演算に伴い，変位計の出力に含まれるノイズ成分はそれぞれ周波数および周波数の二乗に比例する形で速度と加速度の計算結果に現れる．その結果，高周波のノイズ成分は増幅されてしまい，とくに高速な動きが必要なシステムにおいてそのような計測方法では誤差が大きくなり実用的でない場合がある．そこで本章では速度と加速度を他の物理量に変換して計測する手法について説明する．

● 7.1　速度の計測

速度の計測にはドップラー効果に基づくドップラー速度計が一般的に用いられてい

Chapter 7 Measurement of Velocity and Acceleration

Velocity and acceleration are fundamental quantities of motion. Velocity is the distance traveled per unit of time, while acceleration is the rate of change of velocity with respect to time. Velocity and acceleration can thus be calculated from the first- and second-order differentiations of displacement when measured by a displacement sensor, as shown in Chapter 6. However, the high-frequency noise components in the displacement signal will be amplified by the differential operation. As a result, large errors occur in the results, which cannot be employed in systems requiring high-speed motions. This chapter presents the measurement methods by which velocity and acceleration are converted into other quantities for measurement.

● 7.1　Measurement of velocity

The Doppler velocimeter is based on the Doppler effect, which is the phenomenon

る．ドップラー効果とは，音波や光源など波の発生源と観測者との間に相対的な速度が存在する場合，異なる波の周波数が観測者に観測される現象である．

Fig. 7-1(a)と(b)に信号源の移動に伴う波形変化の様子を示す．比較のため，信号源が静止した場合の波形を Fig. 7-1(c)に示す．波を正弦波で表し，信号源から出力される波の周波数を f_s とすると，1秒間に f_s 個の正弦波が出力されるということになる．また，正弦波の周期を T_s とすると，T_s は正弦波の波形が1周期変化するのにかかる時間である．式(7-1)に示すように，T_s は f_s の逆数である．一方，波は信号源から出た後，空間においてある一定の速度で伝搬する．波の伝搬速度は信号源の種類や伝搬空間の媒質によって決まるものであり，信号源の周波数 f_s あるいは周期 T_s には依存しない．なお，波の伝搬速度は伝搬する波の波長と伝搬する波の周期の比，

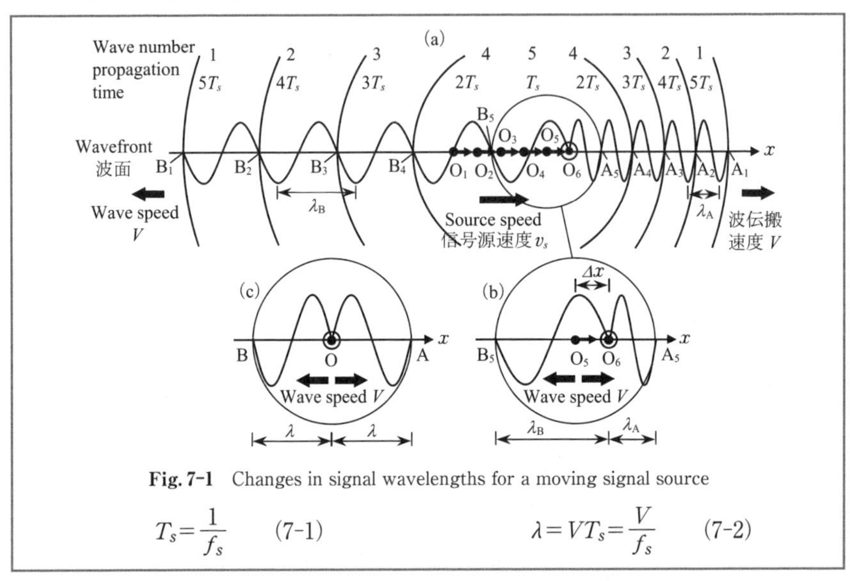

Fig. 7-1　Changes in signal wavelengths for a moving signal source

$$T_s = \frac{1}{f_s} \qquad (7\text{-}1)$$

$$\lambda = VT_s = \frac{V}{f_s} \qquad (7\text{-}2)$$

by which a change in the frequency of a wave can be observed by the observer when it moves relative to the source of the wave.

Figs. 7-1(a) and (b) show the changes in a wave associated with the movement of its source. A wave from a stationary source is shown in Fig. 7-1(c) for comparison. There are f_s sinusoidal waves output from the signal per second. f_s is referred to as the frequency of the wave. T_s, shown in Eq. (7-1), is the period of the wave and is the duration of time of one cycle of the wave. The wave propagates in space at a certain speed after it is output from its source. The propagation velocity is determined by the type of the signal source and the medium in space, but not by the f_s or T_s of the source. It is the ratio of the wavelength of the propagated wave to its period, or the

あるいは波長と伝搬する波の周波数の積となる．なお，波長は正弦波の波形が1周期分変化したとき，波面が空間内で進む距離のことである．ここで注意したいのは，信号の種類と空間の媒質が一定であれば伝搬速度は変化しないが，条件によっては伝搬する波の周波数あるいは周期と波長が変化し得ることである．Fig. 7-1(c)の場合，信号源が静止しているので，空間で伝搬する波は信号源から出力される波の周波数 f_s あるいは周期 T_s を維持したまま伝搬し，伝搬する波の波長 λ が式(7-2)のように求められる．ここで V は波の伝搬速度である．なお，この場合，x 軸の正方向と負方向に伝搬する波の波長は同じである．

Fig. 7-1(a)に，信号源が点 O_1 を出発点にした後，速度 v_S で点 O_2 などを通過して点 O_6 まで移動した瞬間の波形を示す．信号源が点 O_1 にいた時に1番目の波を出力し，点 O_2 に移動したときに2番目の波を出力するように描かれている．Fig. 7-1(a)に示した瞬間では，計5つの波が f_s の周波数で信号源によって出力され，空間で伝搬している．前後2つの波の生成される時間間隔が T_s になる．つまり1番目の波は生成されてから，伝搬速度 V で $5T_s$ の時間だけ伝搬し，5番目の波は生成されてから同じ伝搬速度で T_s の時間だけ伝搬することになっている．

product of the wavelength of the propagated wave to its frequency. Wavelength is defined as the distance needed for the wave to propagate when the waveform changes one period. For a certain signal source and medium in space, the wavelength, the frequency, and the period may change under different conditions while the propagation velocity remains the same. In Fig. 7-1 (c), since the signal source is stationary, the frequency and the period of the propagated wave are f_s and T_s respectively, which are the same as those of the wave output from the signal source. The wavelength λ of the propagated wave is expressed by Eq. (7-2), where V is the propagation velocity of the wave. In this case, the wavelengths of the propagated wave along the positive and negative positions of the x-axis are the same.

Fig. 7-1(a) shows the waveform at the moment when the signal source reaches point O_6 after it starts from point O_1 and passes from O_2 to O_5 at a speed of v_S. The figure is shown so that the signal source outputs the first wave at O_1 and the second wave at O_2. At the moment shown in Fig. 7-1(a), there are five waves that have been output from the source, all of which are propagating in space. The time interval between two adjacent waves is T_s. This means that the first wave and the fifth wave have been propagated over terms of $5T_s$ and T_s respectively, both at the propagation speed V.

Fig. 7-1(b)には，信号源が T_s の時間をかけて点 O_5 から O_6 に距離 Δx だけ移動したときの波形を取り出して示している．移動距離 Δx は式(7-3)のように求められる．この間，左右の波面が x 軸の正方向と負方向に沿ってそれぞれ点 A_5 と点 B_5 まで進み，波形はそれぞれ1周期分変化している．図から，Δx によって x 軸の正方向と負方向に伝搬する波の波長 λ_A と λ_B は異なることが分かり，λ_A と λ_B は式(7-4)と(7-5)のように求めることができる．一方，波の伝搬速度 V は変化しないので，それぞれの方向に伝搬する波の周波数は式(7-6)と(7-7)のように計算される．つまり，信号源の進行方向に沿って伝搬する波は波長が短くなり，周波数が高くなる．それに対して，また信号源の進行方向の反対方向に沿って伝搬する波は波長が長くなり，周波数が低くなる．v_s の正方向を x 軸の正方向に取ると，式(7-4)から式(7-7)までを式(7-8)のようにまとめることができる．f_p と λ_p はそれぞれの波の伝搬周波数と伝搬波長と呼ぶ．

Fig. 7-2には伝搬周波数 f_p と伝搬波長 λ_p の波をレシーバで観察するときの様子を示す．ある時間 Δt の間に波の波面は伝搬速度 V で点Cから点Dまで距離 L_w だけ

$$\Delta x = v_s T_s = \frac{v_s}{f_s} \qquad (7\text{-}3)$$

$$\lambda_A = \lambda - \Delta x = (V - v_s) T_s \qquad (7\text{-}4)$$

$$\lambda_B = \lambda + \Delta x = (V + v_s) T_s \qquad (7\text{-}5)$$

$$f_A = \frac{V}{\lambda_A} = \frac{V}{(V - v_s) T_s} = \frac{V}{V - v_s} f_s \qquad (7\text{-}6)$$

$$f_B = \frac{V}{\lambda_A} = \frac{V}{(V + v_s) T_s} = \frac{V}{V + v_s} f_s \qquad (7\text{-}7)$$

$$f_p = \frac{V}{V \mp v_s} f_s, \quad \lambda_p = \frac{V \mp v_s}{f_s} \qquad (7\text{-}8)$$

Fig. 7-1(b) shows part of the waveform when the source moves over a distance Δx from O_5 to O_6 in a term of T_s. Δx is calculated by Eq. (7-3). The right and left wavefronts move in positive and negative directions of x to A_5 and B_5 respectively, where the waveform changes one period. The wavelengths λ_A in Eq. (7-4) and λ_B in Eq. (7-5) of the propagated waves along the two directions are different from each other. Since V does not change, the corresponding frequencies f_A and f_B of the propagated waves are written in Eqs. (7-6) and (7-7) respectively. The wavelength of the wave propagated along the direction the source is moving becomes shorter and its frequency higher. The wavelength of the wave propagated in the opposite direction, meanwhile, becomes longer and its frequency lower. By taking the positive direction of x as that of v_s, Eqs. (7-4) to (7-7) are summarized in Eq. (7-8), where f_p and λ_p are referred to as the propagation frequency and the propagation wavelength respectively.

Fig. 7-2 shows the wave of f_p and λ_p observed by the receiver moving a distance

移動し，レシーバは速度 v_r で点Ｃから点Ｅまで距離 L_r だけ移動している．レシーバが点Ｃで静止し，波のみが伝搬していると仮定した場合，Δt の間にレシーバに式(7-9)に示される N_1 個の波が通過する．一方，レシーバの移動に伴い，式(7-10)に示す N_2 個の波が余分にレシーバを通過するので，Δt の間にレシーバで観察する波の総数 N は式(7-11)のように N_1 と N_2 の和となる．その結果，レシーバで観察する波の周波数 f_r は式(7-12)のように，波の伝搬周波数 f_p より高くなることが分かる．レシーバが反対方向に移動する場合は，式(7-12)の v_r は負の値になるので，f_r は f_p よ

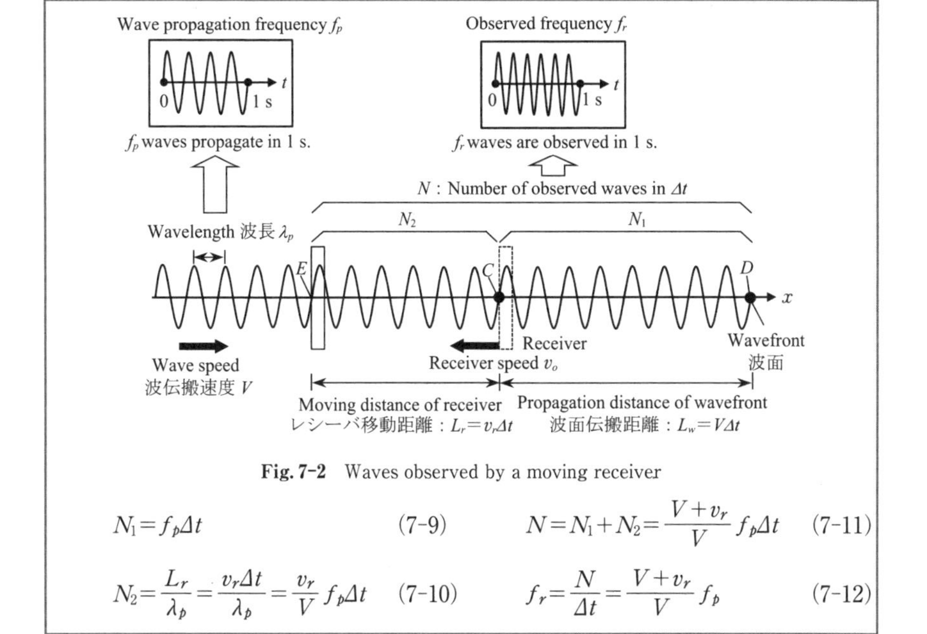

Fig. 7-2　Waves observed by a moving receiver

$$N_1 = f_p \Delta t \qquad (7\text{-}9)$$

$$N = N_1 + N_2 = \frac{V + v_r}{V} f_p \Delta t \qquad (7\text{-}11)$$

$$N_2 = \frac{L_r}{\lambda_p} = \frac{v_r \Delta t}{\lambda_p} = \frac{v_r}{V} f_p \Delta t \qquad (7\text{-}10)$$

$$f_r = \frac{N}{\Delta t} = \frac{V + v_r}{V} f_p \qquad (7\text{-}12)$$

L_r from C to E over term Δt at velocity v_r while the wavefront of the wave moves a distance L_w from C to D at propagation velocity V. If the receiver is kept stationary and only the wave propagates, there will be N_1 waves, as shown in Eq. (7-9), passing through the receiver over the term Δt. When the receiver moves, there will be N_2 more waves, as shown in Eq. (7-10), passing through the receiver and the total number N observed by the receiver during Δt becomes the sum of N_1 and N_2, as shown in Eq. (7-11). Consequently, the frequency f_r of the observed wave shown in Eq. (7-12) becomes higher than f_p. If the receiver moves in the opposite direction, the v_r in Eq. (7-12) takes on a negative value and f_r becomes lower than f_p. When the negative direction of x is taken as the positive direction of v_r, f_r is expressed by Eq.

り低くなる．v_rの正方向をx軸の負方向にとった場合，f_rを式(7-13)のように表すことができる．さらに式(7-8)のf_pを式(7-13)に代入すると信号源とレシーバ両方の移動速度を考慮した場合のf_rは式(7-14)のように得られる．この式から分かるように，信号源あるいはレシーバの移動速度によって信号源の周波数と異なる周波数がレシーバで観察される．これはドップラー効果による現象であり，f_rとf_sとの差Δfをドップラー周波数シフトという．

超音波を利用したドップラー速度計の一例をFig. 7-3に示す．この例では信号源とレシーバが一体化されており，ドップラー周波数シフトΔfがf_sに比べて小さい場合は，移動物体の速度vは式(7-15)のようにΔfにほぼ比例する．超音波のほか，電波と光波もよくドップラー速度計に利用される．生産ラインなどでよく用いられるレーザドップラー速度計の場合は，数十nm/sから数十m/sの速度範囲に対応できるようになっている．また，流体計測に用いられるレーザドップラー流速計は第10章で記述する．

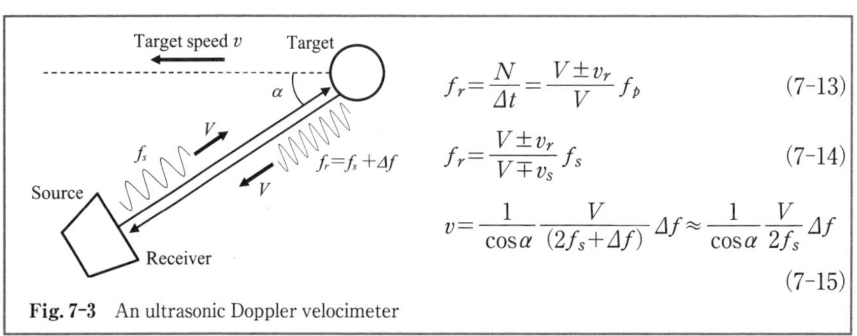

$$f_r=\frac{N}{\Delta t}=\frac{V\pm v_r}{V}f_p \qquad (7\text{-}13)$$

$$f_r=\frac{V\pm v_r}{V\mp v_s}f_s \qquad (7\text{-}14)$$

$$v=\frac{1}{\cos\alpha}\frac{V}{(2f_s+\Delta f)}\Delta f\approx\frac{1}{\cos\alpha}\frac{V}{2f_s}\Delta f \qquad (7\text{-}15)$$

Fig. 7-3　An ultrasonic Doppler velocimeter

(7-13). Eq. (7-14) is obtained by substituting the f_p of Eq. (7-8) into Eq. (7-13). It can be seen that the frequency of the observed wave by the receiver changes with the changes in the velocities of the source and the receiver. This phenomenon is caused by the Doppler effect; the difference Δf between f_r and f_s is referred to as the Doppler shift.

Fig. 7-3 shows an ultrasonic Doppler velocimeter in which the signal source and receiver are combined together. The velocity v of the object is approximately proportional to the Doppler shift Δf if Δf is sufficiently smaller than f_s, as shown in Eq. (7-15). In addition to ultrasonic waves, electrical waves and optical waves can also be employed. The measurement range of a laser Doppler velocimeter typically ranges from tens of nm/s to tens of m/s. The Doppler velocimeter for fluids is described in Chapter 10.

● 7.2 加速度の計測

ターゲット物体の加速度は Fig. 7-4 に示すサイズモ振動系を利用して計測することができる．質量，ばねとダンパから構成される加速度計はねじ止めなどによってターゲットと結合し，一緒に動くようになっている．サイズモ振動系の質量を m，ばね係数を k，ダンパ係数を c とする．物体の変位を $x(t)$ とすると，$x(t)$ の 2 階微分 $\ddot{x}(t)$ が計測対象のターゲット加速度となる．ここで t は時間である．$x_m(t)$ は質量 m の変位を表し，$x_r(t)$ は質量 m とターゲット間の相対変位を表している．Fig. 7-4 に示す原理では，変位センサで相対変位 x_r を計測し，加速度 $\ddot{x}$ と x_r との関係に基づいて，$\ddot{x}$ を x_r から求めるようになっている．

Fig. 7-4 のサイズモ振動系に着目すると，質量 m がダンパとばねからそれぞれ F_c と F_k の力を受け，その運動方程式は式(7-16)と(7-17)で表すことができる．$x_m(t)$，$x(t)$，$x_r(t)$ の関係を示す式(7-18)を式(7-17)に代入して整理すると，式(7-19)のよ

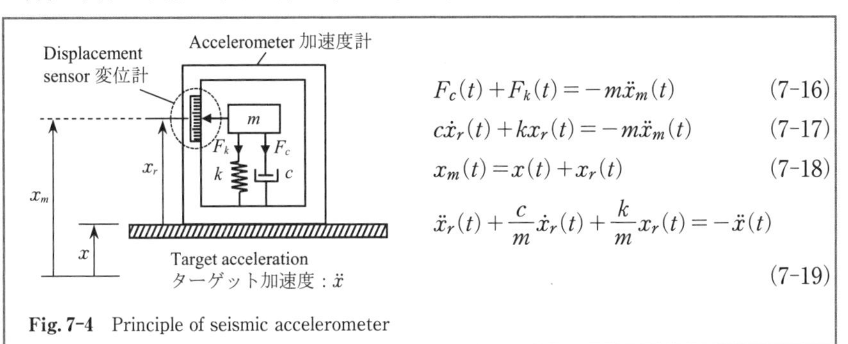

$$F_c(t)+F_k(t)=-m\ddot{x}_m(t) \tag{7-16}$$

$$c\dot{x}_r(t)+kx_r(t)=-m\ddot{x}_m(t) \tag{7-17}$$

$$x_m(t)=x(t)+x_r(t) \tag{7-18}$$

$$\ddot{x}_r(t)+\frac{c}{m}\dot{x}_r(t)+\frac{k}{m}x_r(t)=-\ddot{x}(t) \tag{7-19}$$

Fig. 7-4 Principle of seismic accelerometer

● 7.2 Measurement of acceleration

The acceleration of a target object can be measured by using the seismic vibration system shown in Fig. 7-4. The system consists of a mass m, a spring with a spring constant k, and a damper with a damping ratio c, and is mounted on the target by a screw; it moves with the target. The second-order differentiation $\ddot{x}(t)$ of the displacement of the target $x(t)$ is the acceleration of the target where t is time. The displacement of m is $x_m(t)$, and $x_r(t)$ is the displacement of the mass m relative to the target. To calculate $\ddot{x}$, x_r is employed; x_r is detected by a displacement sensor, based on the relationship between $\ddot{x}$ and x_r.

In Fig. 7-4, the mass m receives the force F_c from the damper and the force F_k from the spring. The equation of motion of the system is shown in Eqs. (7-16) and (7-17). Eq. (7-18) shows the relationship between $x_m(t)$, $x(t)$, and $x_r(t)$. Eq. (7-19), a differentiation equation, is obtained by substituting Eq. (7-18) into Eq. (7-17). The

うに，$\ddot{x}_r$ の項が左辺に，$\ddot{x}$ の項が右辺にそれぞれ含まれる微分方程式が得られる．ターゲットが角振動数 ω で周期的に振動しているとすると，質量 m も同じ角振動数で振動するので，$x_r(t)$ と $\ddot{x}(t)$ を式(7-20)，式(7-21)のようにおくことができる．ここで A と B はそれぞれ $x_r(t)$，$\ddot{x}(t)$ の振幅で，ϕ は $x_r(t)$，$\ddot{x}(t)$ の位相差である．式(7-20)と式(7-21)から分かるように，$x_r(t)$ と $\ddot{x}(t)$ は角振動数の関数である $X_r(j\omega)$ と $\ddot{X}(j\omega)$ とそれぞれ等価である．以下では $X_r(j\omega)$ と $\ddot{X}(j\omega)$ を利用して周波数域で式(7-19)の微分方程式を解く．式(7-20)と式(7-21)を式(7-19)に代入して整理すると，式(7-22)が得られる．さらに機械力学でよく利用される固有角振動数 ω_n（式(7-23)）と減衰比 ζ（式(7-24)）を用いて式(7-22)を式(7-25)のように書き直す．ここでサイズモ振動系の入力を $\ddot{X}(j\omega)$，出力を $X_r(j\omega)$ とすると，系の周波数伝達関数 $G(j\omega)$ は式(7-26)のように得られる．

$$x_r(t)=Ae^{j\omega t}=X_r(j\omega) \quad (7\text{-}20)$$

$$\ddot{x}(t)=Be^{j(\omega t+\phi)}=\ddot{X}(j\omega) \quad (7\text{-}21)$$

$$\left(-\omega^2+j\omega\frac{c}{m}+\frac{k}{m}\right)X_r(j\omega)=-\ddot{X}(j\omega) \quad (7\text{-}22)$$

$$\omega_n=\sqrt{\frac{k}{m}} \quad (7\text{-}23)$$

$$\zeta=\frac{c}{2\sqrt{mk}} \quad (7\text{-}24)$$

$$\left\{-\left(\frac{\omega}{\omega_n}\right)^2+j2\zeta\left(\frac{\omega}{\omega_n}\right)+1\right\}X_r(j\omega)=-\frac{1}{\omega_n^2}\ddot{X}(j\omega) \quad (7\text{-}25)$$

$$G(j\omega)=\frac{X_r(j\omega)}{\ddot{X}(j\omega)}=\frac{\left(-\frac{1}{\omega_n^2}\right)}{1-\left(\frac{\omega}{\omega_n}\right)^2+j2\zeta\left(\frac{\omega}{\omega_n}\right)} \quad (7\text{-}26)$$

left and right sides of Eq. (7-19) contains the terms $\ddot{x}_r$ and $\ddot{x}$ respectively. Assuming the target is vibrating periodically with an angular frequency ω, the mass m also vibrates with the same frequency. Therefore, $x_r(t)$ and $\ddot{x}(t)$ are expressed by Eqs. (7-20) and (7-21), where A and B are the amplitudes of $x_r(t)$ and $\ddot{x}(t)$ respectively. ϕ is the phase difference between $x_r(t)$ and $\ddot{x}(t)$. It can be seen that $x_r(t)$ and $\ddot{x}(t)$ are equivalent to $X_r(j\omega)$ and $\ddot{X}(j\omega)$ respectively, which are used to solve Eq. (7-19) in the frequency domain. Substituting Eqs. (7-20) and (7-21) into Eq. (7-19) gives Eq. (7-22), which is then rewritten as Eq. (7-25) by using the natural angular frequency ω_n from Eq. (7-23) and the damping ratio ζ from Eq. (7-24). The transfer function $G(j\omega)$ of the system is obtained in Eq. (7-26) by taking $\ddot{X}(j\omega)$ as the input and $X_r(j\omega)$ as the output of the system.

The amplitude $|G(j\omega)|$ and phase $\angle G(j\omega)$ of $G(j\omega)$ are evaluated by Eqs. (7-

一方，伝達関数 $G(j\omega)$ は複素数なので，その振幅 $|G(j\omega)|$ および位相 $\angle G(j\omega)$ はそれぞれ式(7-27)，(7-28)のように求めることができる．$|G(j\omega)|$ は相対変位 $x_r(t)$ の振幅 A と加速度 $\ddot{x}(t)$ の振幅 B の比であり，また位相 $\angle G(j\omega)$ は $x_r(t)$ と $\ddot{x}(t)$ の位相差 ϕ である．加速度計に組み込まれる変位センサで計測した質量 m の相対変位 $x_r(t)$ から，式(7-27)，(7-28)に基づいて，ターゲットの加速度 $\ddot{x}(t)$ の大きさおよび位相を定めることができる．

上述のように，加速度計ではサイズモ振動系を用いて，計測量の加速度を変位に変換して計測を行っている．伝達関数 $G(j\omega)$ はその信号変換に伴う感度係数になる．$G(j\omega)$ は原理的に周波数の関数となっているので，第2章で述べたように，そのゲインおよび位相の周波数特性は加速度計の性能にとって重要である．$\omega_n^2|G(j\omega)|$ は無次元の量になるため，ここで便宜上，$20\log(\omega_n^2|G(j\omega)|)$ を加速度計のゲインとして定義し，そのゲインの周波数特性を Fig. 7-5 に示す．図では角振動数比 ω/ω_n を横軸にとっている．ゲインが 0 dB（$\omega_n^2|G(j\omega)|$ が 1）になるのが理想的であるが，ω/ω_n が大きくなるにつれて，ゲインは 0 dB から大きく離れていくことが分かる．ま

$$|G(j\omega)| = \frac{A}{B} = \frac{\dfrac{1}{\omega_n^2}}{\sqrt{\left\{1-\left(\dfrac{\omega}{\omega_n}\right)^2\right\}^2 + \left\{2\zeta\left(\dfrac{\omega}{\omega_n}\right)\right\}^2}} \tag{7-27}$$

$$\angle G(j\omega) = \phi(\omega) = -\tan^{-1}\frac{2\zeta\left(\dfrac{\omega}{\omega_n}\right)}{1-\left(\dfrac{\omega}{\omega_n}\right)^2} \tag{7-28}$$

27) and (7-28) respectively. $|G(j\omega)|$ is the ratio of the amplitude A of $x_r(t)$ to the amplitude B of $\ddot{x}(t)$. $\angle G(j\omega)$ is the difference ϕ between the phases of $x_r(t)$ and $\ddot{x}(t)$. The amplitude and the phase of the acceleration $\ddot{x}(t)$ of the target is thus obtained from Eqs. (7-27) and (7-28), based on the $x_r(t)$ detected by a displacement sensor embedded in the accelerometer.

It can be seen that the acceleration is converted into displacement in the seismic accelerometer. $G(j\omega)$ is the sensitivity coefficient of the signal conversion. Since $G(j\omega)$ is a function of frequency, the gain and phase characteristics of $G(j\omega)$ are important, as addressed in Chapter 2. Based on the fact that $\omega_n^2|G(j\omega)|$ is a dimensionless quantity, the gain of the accelerometer is defined by $20\log(\omega_n^2|G(j\omega)|)$ with a unit of dB for convenience. The frequency characteristics of the gain are shown in Figs. 7-5 and 7-6, where ω/ω_n is taken as the horizontal axis. Ideally, the gain

た，減衰比ζが0や0.3など小さい値をとるときは，振動系はω/ω_nが1の付近で共振し，ゲインが大きく変化するとともに，位相（Fig. 7-6）も激しく変わるので，共振点付近では不安定になる可能性がある．一方，ζが$\sqrt{2}/2$（約0.7）以上になると振動系は共振しないことになるため，多くのサイズモ加速度計では，ζが0.7付近の値をとるように振動系の質量m，ばね係数k，ダンパ係数cを設計している．なお，ζが$\sqrt{2}/2$のときに，加速度計の遮断周波数ω_cはちょうど振動系の固有周波数ω_nになる．

Fig. 7-7に，変位センサとして第6章で述べたひずみゲージあるいは静電容量型変位計を利用した加速度計の概略図を示す．ひずみゲージ加速度計では板ばねはサイズモ振動系のばねの働きをするとともに，質量の相対変位をばねのひずみに変換してひずみゲージで検出する役割も果たしている．また，一般的に粘度変化の少ないシリコンオイルをダンパに採用している．市販のひずみゲージ式加速度計はDCから数

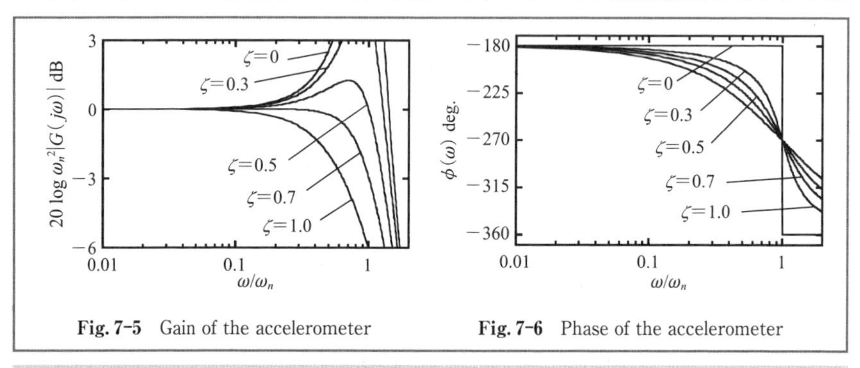

Fig. 7-5　Gain of the accelerometer

Fig. 7-6　Phase of the accelerometer

should be 0 dB where $\omega_n^2\,|G(j\omega)|$ is 1 for all frequencies. However, it will deviate from 0 dB when ω/ω_n gets larger. When ζ is small, such as 0 and 0.3 in Figs. 7-5 and 7-6, the system resonates with significant changes in the gain and the phase when ω/ω_n is close to 1, which makes the system unstable. When ζ is larger than $\sqrt{2}/2$ (approximately 0.7), the system will no longer resonate. Many accelerometers are designed to have a ζ of approximately 0.7 by selecting proper values of m, k, and c, where the cutoff frequency of the accelerometer ω_c is equal to ω_n.

Fig. 7-7 shows an accelerometer using the strain gauge or the capacitive displacement sensor described in Chapter 6. For the strain gauge accelerometer, the cantilever works as the spring of the vibration system and also converts the displacement of the mass into the strain to be detected by the strain gauge. Silicon oil with a stable coefficient of viscosity is typically employed as the damper in the accelerometer. The bandwidth of a commercial accelerometer is in DC and runs to kHz までの周波数帯域を持ち，加速度測定範囲は最大で 10 km/s^2 程度になっている．一方，静電容量型加速度計はマイクロマシンの技術で大量生産できる．MEMS 加速度計とも呼ばれるこの加速度計は周波数帯域と加速度測定範囲はひずみゲージ式とほぼ同じであるが，小型で安価な特徴があり，自動車やスマートフォンなどに応用されている．

そのほか，変位センサの代わりに第 8 章で述べる圧電型力センサを用いて質量 m が受ける力を計測し，加速度を求める加速度計もある（Fig. 7-8）．質量はねじなどで力センサの圧電素子と結合されている．圧電型加速度計では通常ダンパを使用しないシンプルな構造をとっている．この場合，サイズモ振動系の減衰比 ζ はほぼゼロとなり，Fig. 7-5 と 7-6 から分かるように，サイズモ振動系の固有角振動数 ω_n より十分に小さい角振動数の範囲でしか安定した加速度計測ができない．一方，圧電素子は高剛性でばね係数 k が極めて大きいため，ω_n が非常に大きい特徴がある．市販の圧

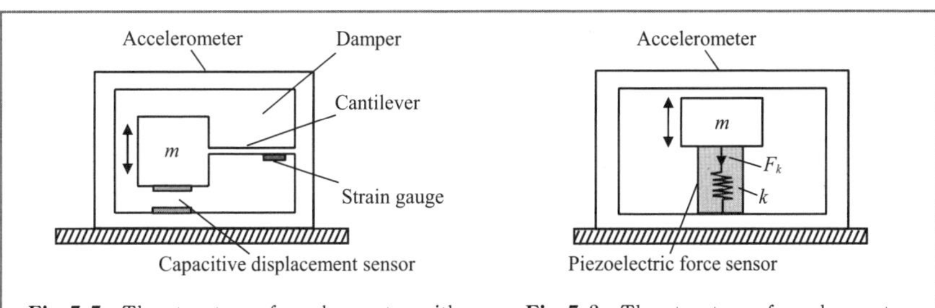

Fig. 7-7 The structure of accelerometer with a strain gauge or capacitive displacement sensor

Fig. 7-8 The structure of accelerometer with a piezoelectric force sensor

several kHz. The measurement range of acceleration can be up to 10 km/s^2. The capacitive type accelerometers that can be mass-produced by the silicon-based MEMS technology have the advantages of compactness and low-cost, and are widely used in automobiles and smartphones. The bandwidth and measurement range of the MEMS accelerometer are comparable to those of the strain gauge accelerometer.

There are also accelerometers that use a piezoelectric force sensor to measure the force applied to the mass m, as shown in Fig. 7-8, where the mass and the force sensor are connected with each other. The damping ratio of this type of accelerometer is almost zero, since no damper is employed. Stable measurements can only be made when the frequency is much lower than ω_n, as shown in Figs. 7-5 and 7-6. On the other hand, the spring constant k of the piezoelectric force sensor is extremely high, resulting in a large ω_n. For this reason, a commercial piezoelectric accelerometer has a bandwidth of several tens of kHz and a measurement range up to 1000 km/s^2 with a

電加速度計では，数十 kHz の周波数帯域を持ち，加速度測定範囲は最大で 1000 km/s^2 程度になっている．構造的に機械強度が高いこともあり，振動や衝撃計測などによく用いられる．ただし，第8章で述べる圧電力センサの特性上，加速度の直流成分が測定しにくい欠点がある．

【演習問題】

7-1) 波の周波数 f，波長 λ，周期 T のうち，伝搬空間の媒質によって変わるのはどれか．

7-2) 周波数 20 kHz，音速 340 m/s の音波が出力される静止音源に向かって，60 km/h の速度で近付いてきている自動車の中で観測される周波数 f_r を求めよ．

7-3) Fig. 7-3 において，設置誤差によって角 α に標準不確かさ u_α が生じた場合の，それに対応する物体速度 v の計測結果の標準不確かさ u_v を求めよ．

7-4) Fig. 7-4 において，x を入力，x_r を出力としたときの伝達関数 $G_x(j\omega)$ を求めよ．

7-5) 減衰比 ξ が 0.5 のサイズモ加速度計の遮断周波数 ω_c を求めよ

strong mechanical structure, which is suitable for use in the measurement of vibrations and shocks. However, it is difficult to measure the DC component because of the characteristics of the piezoelectric force sensor, as described in Chapter 8.

【Problems】

7-1) Which does not change with the change of the medium in the propagation path: the frequency f, the wavelength λ, or the period T of the wave?

7-2) Calculate the observed wave frequency f_r in a car moving 60 km/h toward a stationary sound source with a signal frequency of 20 kHz and a signal speed of 340 m/s.

7-3) Evaluate the standard uncertainty u_v of the measurement result of the target velocity v when a setting error occurs in α with a standard uncertainty u_α in Fig. 7-3.

7-4) Obtain the transfer function $G_x(j\omega)$ by taking x and x_r from Fig. 7-4 as the input and output respectively.

7-5) Show the cutoff frequency ω_c of a seismic accelerometer for a damping ratio ξ of 0.5.

第8章　力と質量の計測

ロボットなどの各種システムでは電気などのエネルギーをモータなどの機構を介して力に変換し，その力で目的に応じた仕事をすることになっている．また，力を中心に扱う各種力学は物理学，工学など多くの学問分野の礎をなしているといえる．力とは，物体の速度を変化させる働きのこと，また物体の形を変形させる働きのことである．本章では，後者を利用して力を計測する力センサについて述べる．また，重力の計測を基本とする質量の計測法についても説明する．

● 8.1　力の計測

力の計測に用いられる力センサ（ロードセル）は，弾性体の変形を利用するものと，圧電材料の圧電効果を利用するものに分類することができる．

弾性体を利用した力計測　この方式では，計測量の力を弾性体に加えたときに生じる

Chapter 8　Measurement of Force and Mass

In a system such as a robot, energy is converted into forces by using elements such as motors to accomplish a task. Force is a basic quantity in mechanics that forms the foundation of physics and engineering. A force can change the velocity of an object. It can also deform an object. In this chapter, the latter is employed to measure of force. The measurement of mass based on the force of gravity is also described.

● 8.1　Measurement of force

Force sensors (load cells) are based on the deformation of an elastic body or on the piezoelectric effect of a piezoelectric material.

Elastic body-based force measurement　In this method, force measurement

弾性体変形量を検出することによって，力の計測を行うことになっている．弾性変形量の検出にはひずみゲージがよく用いられる．

ひずみゲージロードセルは弾性体の形によって，コラム型とベンディングビーム型に分類される．Fig. 8-1 に長方体や円柱体などのコラムを弾性体に用いるコラム型ロードセルの原理を示す．xy 断面積が S のコラムの中心軸（z 軸）に沿って力 F を印加すると，z 軸方向に圧縮応力 σ（式(8-1)）が生じ，それに応じてコラムが z 軸方向に縮み，x および y 方向に伸びるように変形する．コラムのヤング率（縦弾性係数）を E，ポアソン比を ν とすると，z 方向および x 方向に生じるコラム表面のひずみ ε_z と ε_x はそれぞれ式(8-2)，(8-3)のとおり表すことができる．初期抵抗値 R，ゲージ率 k_s のひずみゲージ G1 と G2 をコラムに張り付けて，ε_z と ε_x をそれぞれ計測すると，G1 と G2 の抵抗値にそれぞれ ΔR_1 と ΔR_2 の変化量が生じる．第 6 章で述べたひずみゲージの原理より，式(8-4)と(8-5)が得られ，それらを整理すると式(8-6)にな

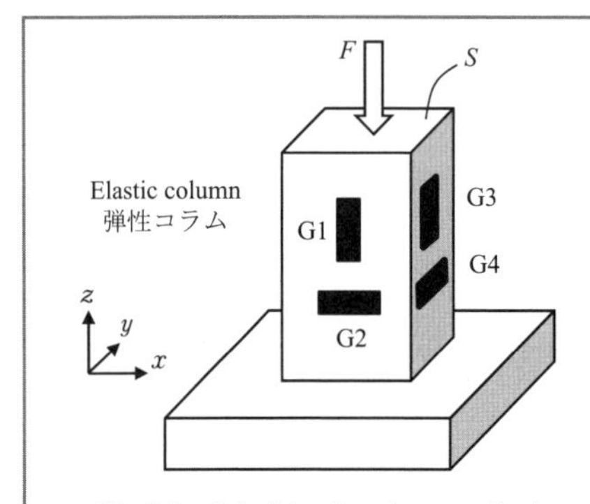

Fig. 8-1 Principle of strain gauge load cell: column type

$$\sigma=\frac{F}{S} \tag{8-1}$$

$$\varepsilon_z=\frac{\sigma}{E}=\frac{F}{ES} \tag{8-2}$$

$$\varepsilon_x=-\nu\varepsilon_z=-\frac{\nu F}{ES} \tag{8-3}$$

$$\frac{\Delta R_1}{R}=k_s\varepsilon_z=k_s\frac{F}{ES} \tag{8-4}$$

$$\frac{\Delta R_2}{R}=k_s\varepsilon_x=-k_s\frac{\nu F}{ES} \tag{8-5}$$

$$F=\frac{ES}{k_sR}\Delta R_1=-\frac{ES}{\nu k_sR}\Delta R_2 \tag{8-6}$$

is made through detecting the deformation of an elastic body. The deformation detection is often made by using strain gauges.

Strain gauge load cells are classified into the column type and the bending beam type. Fig. 8-1 shows the column type, in which a cuboid or a cylinder is employed as the elastic body. When a z-directional force F is applied to a column with an xy cross-sectional area S, the compressive stress σ in Eq. (8-1) will be caused in z, making the column compress in the z direction and stretch in the x and y directions. Denoting the Young's modulus (modulus of longitudinal elasticity) and the Poisson's ratio of the column by E and υ respectively, the column surface strain ε_z in z and ε_x in x are expressed in Eqs. (8-2) and (8-3) respectively. Based on Chapter 6, the resistance change ratios of the strain gauges G1 and G2 attached on the column surface for

る．このように，ひずみゲージの抵抗出力 ΔR_1 あるいは ΔR_2 から力 F を求めることができる．

実際のコラム型ロードセルの場合，Fig. 8-1 に示すように，さらに一組のひずみゲージ G3 と G4 を追加し，4 つのひずみゲージで第 14 章のブリッジ回路を差動的に構成することによって，ひずみゲージの温度ドリフトの影響を低減するとともに，力計測の感度を向上させるようにしている．また，測定軸（z 軸）以外からの力の影響を低減するように，ダイヤフラムの機構を追加するなどの工夫がされている（Fig. 8-2)．

コラム型は計測感度が低いものの，構造が単純で小型のサイズでも大きな測定範囲が得られる特徴がある．また，複数コラムの採用によって，測定範囲をさらに拡大することが可能である．市販のコラム型ロードセルの一般的な測定範囲は数 kN から数 MN になっている．一方，測定範囲は限定され，剛性も低いが，計測感度が高い特徴

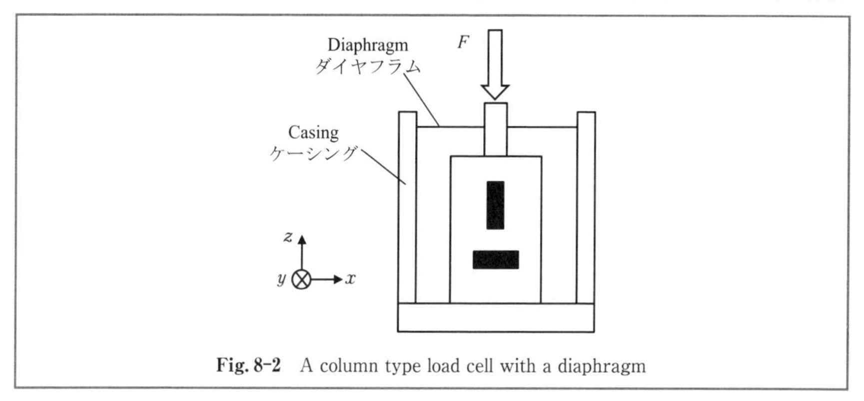

Fig. 8-2　A column type load cell with a diaphragm

detecting ε_z and ε_x can be written in Eqs. (8-4) and (8-5) respectively, where R and k_s are the initial resistance and the gauge factor of the strain gauge. F can then be measured from the resistance change ΔR_1 or ΔR_2, based on Eq. (8-6).

For a practical column load cell, as shown in Fig. 8-1, another pair of strain gauges G3 and G4 is added so that four strain gauges can be employed in the bridge circuit shown in Chapter 14 with a differential structure for thermal drift compensation and force sensitivity improvement. A diaphragm is also often added to reduce the influence of the force component in a direction other than z, as shown in Fig. 8-2.

Although the sensitivity of the column type is low, it can achieve a large range with a simple structure and a compact size. The range can be further expanded by using multiple columns. The range of a commercial column load cell runs from several kN to several MN. In contrast, a bending beam load cell can achieve a higher sensitivity, but its range is limited and its stiffness low.

を持つのが Fig. 8-3 に示すベンディングビーム型である．

Fig. 8-3 に，ベンディングビームとして最もシンプルな片持はりを用いるベンディングビーム型ロードセルの原理を示す．ひずみゲージ G1 がはりの上面に，G2 がはりの同じ位置の下面にそれぞれ接着されている．z 軸方向に沿って片持はりの自由端付近に力 F を印加すると，はりが xz 面内でたわむようになる．はりの yz 断面は厚さ h，幅 b の長方形であるとすると，その断面二次モーメント I は式(8-7)のようになる．図に示すはりの網掛け部分は曲げモーメント M を受ける．力 F の作用点からひずみゲージの中心までの距離を L とし，L がひずみゲージの長さに比べて十分に大きいとすると，M は式(8-8)に求めることができる．M の作用を受けて，はりの上面が伸び，下面が縮むので，G1 と G2 で測定するはりの表面のひずみ ε_1 と ε_2 は式(8-9)のように大きさは同じであるが，符号が逆となる．F の大きさは式(8-10)のようにひずみゲージ抵抗変化量の大きさ ΔR から求めることができる．

コラム型と同様に，Fig. 8-4 に示すように，実用的なベンディングビーム型ロード

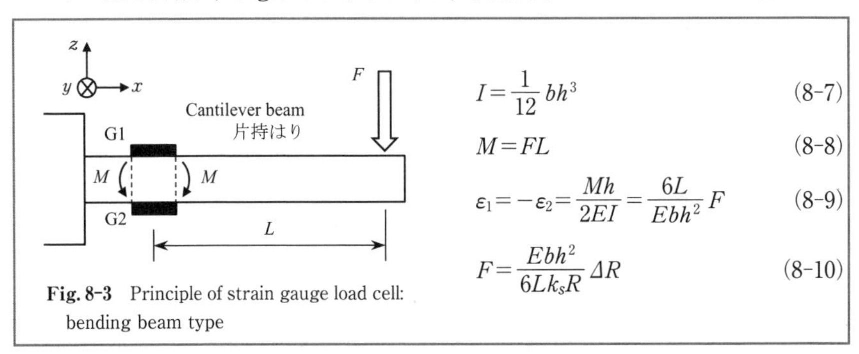

$$I=\frac{1}{12}bh^3 \tag{8-7}$$

$$M=FL \tag{8-8}$$

$$\varepsilon_1=-\varepsilon_2=\frac{Mh}{2EI}=\frac{6L}{Ebh^2}F \tag{8-9}$$

$$F=\frac{Ebh^2}{6Lk_sR}\Delta R \tag{8-10}$$

Fig. 8-3　Principle of strain gauge load cell: bending beam type

Fig. 8-3 shows the principle of the simplest cantilever bending beam type. Strain gauges G1 and G2 are glued to the upper and lower surfaces of the beam, respectively. When a force F is applied to the free end of the beam along the z direction, the beam is bent in the xz plane. The second moment of area I of the rectangular section of the beam in the yz plane with a thickness h and a width b is expressed in Eq. (8-7). The hatched part of the beam in the figure is subject to the bending moment M in Eq. (8-8). Since the upper and the lower parts of the beam are stretched and compressed respectively, the surface strains ε_1 and ε_2 measured by G1 and G2 have the same amplitude but with opposite signs, as Eq. (8-9) shows. The magnitude of F can thus be measured from that of the resistance change ΔR, based on Eq. (8-10).

As with the column type, four strain gauges are employed in a practical bending beam type, as shown in Fig. 8-4, for a differential bridge circuit. However, in a

セルでは，4つのひずみゲージを設置して，ブリッジ回路を差動的に構成している．ただし，片持はりを利用した構造では，力 F の作用点位置が変化すると，Fig. 8-3 の L が変化する．式(8-10)から分かるように，L の変化によって力の測定値に誤差が生じてしまう．この問題を避けるために，実用上は2個のベンディングビームを用いるダブルビーム型が一般的である．Fig. 8-5 に示すように，ダブルビーム構造では，はり1と2の両端はそれぞれ剛体部AとBに拘束されている．剛体部Bを介して印加される力 F によって，はり1と2は図に示すようにS字変形し，はりの x 方向中心線 CC′ はS字の中心を通過する．CC′ を中心に対称な位置にあるひずみゲージ G1 と G2，また G3 と G4 の出力は大きさは同じである．なお，ダブルビーム型では，ひずみゲージで検出するひずみの大きさは，ひずみゲージの位置から CC′ までの距離 $L/2$ にのみ依存するので，剛体部B上での力 F の作用点が変化してもひずみゲージの出力が変わらず高精度に F を計測できるという特徴がある．

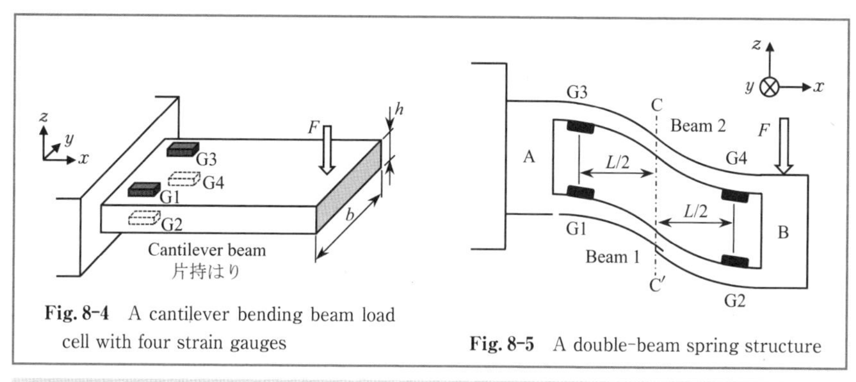

Fig. 8-4　A cantilever bending beam load cell with four strain gauges

Fig. 8-5　A double-beam spring structure

cantilever beam, a change in the application point of F will cause a change in L of Fig. 8-3, which will introduce a measurement error in F. A double-beam structure is thus employed in the practical bending beam load cell shown in Fig. 8-5. The two ends of beams 1 and 2 are constrained by rigid parts A and B, respectively. When the force F is applied to part B, the beams show an S-shaped deformation whose center is located on the beams' center line CC′. The outputs of G1 and G2 and those of G3 and G4, which are located in symmetrical positions about CC′, have the same amplitudes and opposite signs. The output of the strain gauge depends only on its distance $L/2$ to CC′. Accurate force measurement can thus be carried out since the output will not change with the change of the application position of F on part B.

Similarly, surface strains will be caused on a shaft when the shaft is subject to a torque (torsional moment) T. Consider the deformation at point P on the cylindrical

同様に，動力装置の伝導軸などに加えられたトルク（ねじりモーメント）によって軸表面にひずみが生じることを利用してトルクを計測することができる．Fig. 8-6 に示すように，一様な円形断面を持つ丸棒の軸に対し，軸中心回りにトルク T が作用するとして，軸表面上の点Pにおける変形について考える．軸の直径を D とすると，軸の断面二次極モーメント I_P は式(8-11)となる．T によって円周方向に沿ってせん断応力 τ が発生するが，円形断面の場合，軸の断面はねじられた後も yz 平面内で同じ直径の円形を保つので，軸表面の軸方向（AA′ 方向）および円周方向には垂直応力が生じないことが分かる．この純せん断の応力状態では，AA′ から 45° 傾斜した BB′ と CC′ 方向にはそれぞれ τ と同じ大きさの引張り応力（σ）と圧縮応力（$-\sigma$）が生じる．軸の横弾性係数を G，ヤング率を E，ポアソン比を ν すると，τ と σ を式(8-12)から得ることができる．また，対応する表面ひずみ $\varepsilon_{BB'}$ と $\varepsilon_{CC'}$ を式(8-13)のように表すことができる．BB′ あるいは CC′ 方向に沿って張り付けた G1 と G2 など

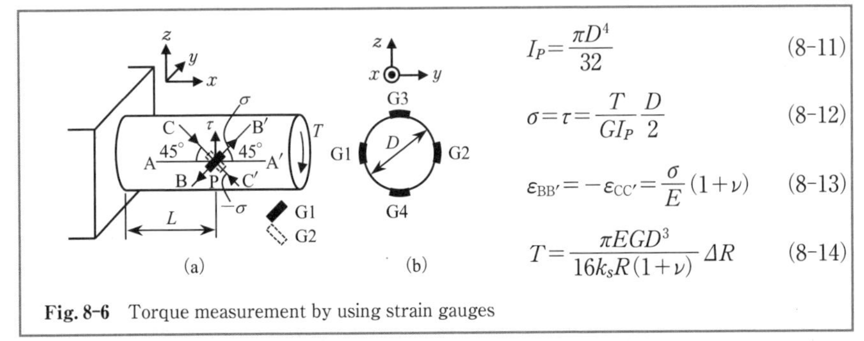

$$I_P = \frac{\pi D^4}{32} \quad (8\text{-}11)$$

$$\sigma = \tau = \frac{T}{GI_P}\frac{D}{2} \quad (8\text{-}12)$$

$$\varepsilon_{BB'} = -\varepsilon_{CC'} = \frac{\sigma}{E}(1+\nu) \quad (8\text{-}13)$$

$$T = \frac{\pi EGD^3}{16k_sR(1+\nu)}\Delta R \quad (8\text{-}14)$$

Fig. 8-6　Torque measurement by using strain gauges

shaft in Fig. 8-6. The polar moment of inertia of area I_P of the shaft with a uniform round section of diameter D is expressed in Eq. (8-11). A shear stress τ is caused along the circumference by T. However, since the round shape of the shaft section and its diameter D are maintained, there are no normal stresses along the AA′ direction and the circumference. In this pure shear stress state, there will be compressive stress (σ) and tensile stress ($-\sigma$) along BB′ and CC′ with an angle of 45° to AA′. The amplitudes of σ and τ, which are the same, and the corresponding surface strains $\varepsilon_{BB'}$ and $\varepsilon_{CC'}$ are expressed in Eqs. (8-12) and (8-13) respectively, where G, E, and ν are the modulus of transverse elasticity, the Young's modulus, and the Poisson's ratio of the shaft respectively. T can thus be evaluated from ΔR of the strain gauges G1 and G2 aligned along the BB′ or CC′, based on Eq. (8-14). A differential bridge circuit can be constructed by using the four strain gauges shown in Fig. 8-6(b).

のひずみゲージ抵抗変化量 ΔR から，T の大きさは式(8-14)のように求めることができる．なお，Fig. 8-6(b)に示す4つのひずみゲージを利用して差動型ブリッジ回路を構成することができる．

圧電効果を利用した力計測　水晶や圧電セラミックスなどの圧電体に力を加えると，機械的な微小変形に応じて表面電荷が同時に生じるという圧電効果が現れる．ペロブスカイト型結晶構造を持つ強誘電体圧電セラミックスを圧電体として利用した力検出の原理を Fig. 8-7 に示す．組成式が ABO_3 で表されるこの結晶は図に示すように A, B, O（酸素）イオンからなる．例えば PZT と呼ばれる $Pb(Zr, Ti)O_3$ では Pb が A イオン，Zr/Ti が B イオンとなる．

このような圧電体は，キュリー温度以上の常誘電相では，B イオンが中心にある立方晶系であり，自発分極を持たない．キュリー温度以下の強誘電相では正方晶系が形成され，B イオンが結晶内部で中心から δ_0 だけ上方向にずれた位置に移動する（Fig. 8-7(a)）．それによって，圧電体内部で双極子が誘起されて自発分極が発生し，圧電体の上下表面付近の圧電体内部の部分ではそれぞれ正電荷と負電荷が生じる．ただ

Piezoelectric effect-based force measurement　When a force is applied to a piezoelectric material, surface electric charges will accumulate due to the piezoelectric effect. Fig. 8-7 shows the principle of measuring force by using ferroelectric and piezoelectric ceramics with a Perovskite crystal structure having the general composition ABO_3 composed of A, B and O (oxygen) ions. For $Pb(Zr, Ti)O_3$, known as PZT, Pb is the A ion and Zr/Ti is the B ion.

Such a piezoelectric material has a cubic crystal structure with the B ion at the center. It loses its spontaneous polarization in the paraelectric phase above the Curie temperature. In the ferroelectric phase below the Curie temperature, the material has a tetragonal crystal structure where the B ion moves a distance δ_0 from the center position (Fig. 8-7 (a)). Dipoles are thus induced and spontaneous polarization appears, resulting in positive and negative internal charges underneath the upper and the lower surfaces of the piezoelectric material. Since the internal charges are typically neutralized by the floating charges in the air that are attracted to the surfaces, the total charge amount on each of the upper and the lower surfaces is zero.

As shown in Fig. 8-7(b), when a force F is applied along z, the B ion moves toward the center due to compressive stress. Since the distance δ_1 is smaller than δ_0,

し，通常は空気中の浮遊電荷などが圧電体上下表面に引き寄せつけられてこれらの内部電荷と中和するため，上下それぞれの表面の総電荷量がゼロになる．

そこで z 軸に沿って力 F を加えると，Fig. 8-7(b)に示すように，圧縮応力によって結晶が変形し，結晶中心からのBイオン位置のずれ量 δ_1 が δ_0 に比べて小さくなるため，自発分極が減り，圧電体の上下表面付近の内部電荷量が減少する．その結果，表面の外部電荷の量が内部電荷量より上回り，圧電体の上下表面にはそれぞれ総電荷量 Q を持つ．正圧電効果とも呼ばれるこの現象は力の計測に利用され，高感度と高剛性の特徴がある．なお，圧電体に外部電場を加えると変形が生じるという逆圧電効果もあり，振動子やアクチュエータに利用されている．

一方，説明の都合上，図では1個の結晶だけを示しているが，実際の圧電体は，無数の結晶の集合体である．近くの双極は，ワイス・ドメインという微小領域で同じ方

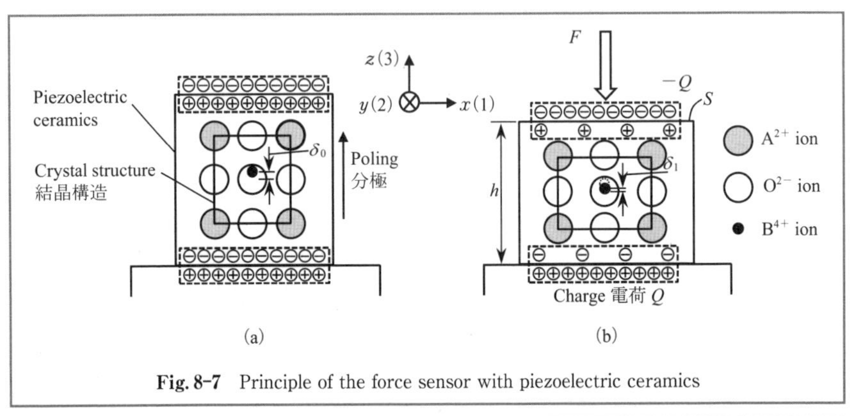

Fig. 8-7 Principle of the force sensor with piezoelectric ceramics

the internal charges underneath the surfaces decrease with the decrease in spontaneous polarization. As a result, the external charges on the surfaces are greater than the internal charges, generating a total amount Q of surface charges. This piezoelectric effect or direct piezoelectric effect is utilized for force measurement with high sensitivity and high stiffness. The material will also be deformed when it is subject to an external electric field. This inverse piezoelectric effect is used in vibrators and actuators.

Although only one crystal is shown in the figures for clarity, a piezoelectric material is composed of many crystals. Dipoles near each other are aligned in a micro-region called Weiss domains. However, since the orientations of the domains are random, the piezoelectric effect cannot be obtained for the entire material. In what is called the poling process, a strong electric field of several V/m is then applied across

向に並ぶ．しかし，各ドメインの配置はランダムであるため，圧電体全体としては圧電効果が得られない．そこで圧電体に数 V/m の強い電界を加えて，ドメインの方位を揃えるという分極操作を行う．圧電効果を取り扱う圧電方程式では，テンソル表記法が採用されており，分極方向は通常 3 軸方向（Fig. 8-7 では z 軸方向）に定義されている．圧電体に外部電場が作用せず，力 F のみ作用する場合は，圧電方程式が式(8-15)となる．電荷量 Q を第 14 章のチャージアンプで検出すれば，力 F を求めることができる．なお，d_{33} は圧電定数といい，圧電体の材料で決まる定数である．また，圧電体を用いた力センサでは，センサの線形性や剛性などを向上させるために，通常圧電体に予荷重を加えている．

圧電体は，静電容量が C_d のコンデンサと抵抗値が R の絶縁抵抗を並列に接続した等価回路で表現できる．コンデンサの放電で力センサの電荷出力 q は式(8-16)のように時間 t の関数となり，圧電体に加わる力が一定に保持されていても，q は減少していく．そのため，圧電型力センサは静的な計測には向かない．なお，C_d は式(8-17)のように圧電体の厚さ h，断面積 S，誘電率 ε_{33} から計算でき，また R は h と S および圧電体の抵抗率 ρ から計算できる．RC_d は圧電素子の時定数 τ と呼ばれる．ただし，センサの時定数は後続のチャージアンプにも影響されるので，それより短くなる．

$$F=\frac{Q}{d_{33}} \quad (8\text{-}15) \qquad q=Qe^{-\frac{t}{RC_d}} \quad (8\text{-}16) \qquad C_d=\varepsilon_{33}\frac{S}{h} \quad (8\text{-}17)$$

the material to align the orientations. In the piezoelectric equations expressed in tensor form, the poling direction is denoted as the "3" axis (the z axis in Fig. 8-7). Eq. (8-15) is obtained when the force F is applied. F can be evaluated by detecting the charge Q with the charge amplifier shown in Chapter 14. d_{33} is the piezoelectric coefficient of the material. The piezoelectric material is typically pre-loaded in the force sensor for improvement of sensitivity and stiffness.

A piezoelectric material can be expressed by an equivalent circuit composed of a capacitor C_d and an insulation resistor R connected in parallel. The charge output q of a force sensor is expressed in Eq. (8-16) as a function of time t, based on the discharge of the capacitor. q will decrease even when F is kept constant. For this reason, a piezoelectric force sensor is not suitable for static force measurement. C_d is evaluated in Eq. (8-17) from the thickness h, the cross-sectional area S, and the permittivity ε_{33} of the piezoelectric material. R can be evaluated from h, S, and the resistivity ρ of the material. RC_d is called the time constant τ of the material. τ will be shortened when the material is connected to a charge amplifier.

● 8.2　質量の計測

質量とは，物体の動かしにくさの度合いを表す量のことであり，その物体固有の性質を示すものである．それに対して，日常生活でよく用いる重さあるいは重量とはその物体に働く重力の大きさのことであり，質量と重力加速度の積で定義されている．質量は不変であるが，同じ物体でも重力加速度が異なる場所ではその重さが変わってしまう．

一方，質量は直接計測の手段がなく，物体に働く重力を介して間接的に計測するのが一般的である．例えばFig. 8-8に示す左右のアームの長さが同じ機械式天秤を使って物体の質量 m を測る場合は，左の皿に物体を載せておき，目盛板における指針の位置を観察しながら天秤の左右の部分が釣り合うという平衡状態になるまで右の皿に質量が既知の分銅を載せていく．平衡状態においては，物体と分銅に作用する重力 W_m と W_M は同じになるので，重力加速度の影響を受けずに分銅の総質量 M から物体の質量 m を知ることができる．このように，計測量と基準量を比較して両者が等しくなる平衡状態を検知し，その時の基準量から計測量を求める計測法を零位法と呼ぶ．平衡検知の必要範囲が狭いので，高い感度と精度で平衡が検知できる．そのため，零位法は基準量の精度に近い精度での計測ができる特徴がある．しかし，この方

● 8.2　Measurement of mass

Mass is the measure of the resistance of an object to a change in its state of motion. Mass is not the same as "weight", a term more often used in daily life. Weight is the force of gravity acting on an object and is a product of mass and the acceleration of gravity. The weight of an object can change at different places with different accelerations of gravity, while its mass is always the same.

Mass is measured by measuring weight. In the measurement of a mass m by a mechanical balance scale with equal arms, the mass is placed in the left pan and standard masses of known weight are added to the right pan until the beam reaches equilibrium, with the readings observed by the naked eye. Because the forces of gravity W_m and W_M acting on the mass and standard weights are the same in equilibrium, m can be obtained from the total mass M of the standard weights without the influence of the acceleration of gravity. In the null method employed here, a standard quantity is adjusted in a balancing mechanism to equilibrium, where it is the same as the unknown measurand's quantity. The latter can thus be identified with the value of the former. High-accuracy and high-sensitivity detection of equilibrium can

法で質量を測る場合は，用意できる分銅に限りがあることに加えて，手動で平衡をとる手間と時間がかかる問題がある．

そのため，現在では精密質量計測には Fig. 8-9 に示す電磁力平衡式電子天秤が主流になっている．同じ零位法が採用されているが，磁場中に置かれるコイルに電流 I を流して発生した力 F を分銅の代わりに用いる．変位センサを使って平衡を検知しながら，電流 I をフィードバック制御して W_m と F の平衡をとる．事前に分銅を使って校正しておいた電流 I と質量 m との関係に基づいて，物体の質量 m を平衡状態の電流 I から求めることができる．力 F に比例する電流 I を精密に計測制御することができるので，電子天秤は非常に高い計測感度が実現できる．ただし，精密な質量計測の場合は，温度や湿度，気流，静電気，振動など測定環境の影響を受けるので対策が必要となる．また，設置時に電子天秤本体を水平にすることも重要である．

一方，ひずみゲージロードセルを用いた電気式はかりもある．物体の重量に比例す

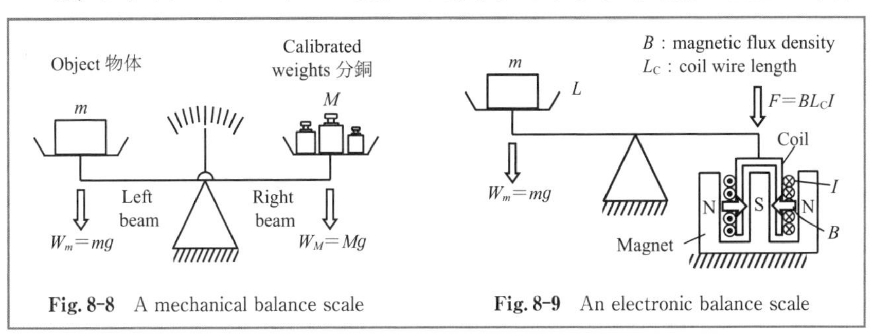

Fig. 8-8　A mechanical balance scale

Fig. 8-9　An electronic balance scale

be achieved since the range necessary for the detection is small. The measurand can thus be measured with an accuracy close to that of the standard quantity. However, for a mechanical balance, only limited standard weights are available, while the manual operation of creating equilibrium is time-consuming.

The electronic balance scale with the null method shown in Fig. 8-9 is now widely used for precision mass measurement. The electromagnetic force F generated by applying a current I to a coil placed in a magnetic field is employed instead of standard weights. The equilibrium between W_m and F is reached by feedback controlling I while monitoring the equilibrium with a displacement sensor. The mass m can be evaluated from I at equilibrium based on the relationship of I and m, which is calibrated by using standard weights beforehand. High sensitivity can be achieved with the precision measurement and control of I that is proportional to F. It is important to pay attention to the temperature, humidity, air flow, static electricity, and vibration, along with the leveling of the balance.

る弾性体の変形という偏位量から質量を求めるという偏位法の原理を採用しているこの方式は，零位法の電磁力平衡式に比べて精度が劣るものの，フィードバック制御機構を必要としないため，測定速度が速く，構造が簡単という特徴がある．またコラム型ロードセルを用いれば重量物の計測にも対応できる．この方式も重力を計測しているので，質量を求めるためには事前に分銅による校正が必要となる．

【演習問題】

8-1）Fig. 8-1 において，ヤング率 E，断面積が 100 mm^2 のステンレス鋼コラムに 100 N の力が加わったとき，E を調べたうえで，G1（k_s=2）の $\Delta R_1/R_1$ を求めよ．

8-2）Fig. 8-5 の力 F とひずみゲージ G1 の出力抵抗値 ΔR の関係を示せ．なお，計算に必要な諸元はビームの長さ以外は Fig. 8-3 のものと同じである．

8-3）圧電体の時定数 RC_d は圧電体の誘電率 ε_{33} および抵抗率 ρ にのみ依存することを証明せよ．

8-4）質量 5 kg の物体がそれぞれ東京とロンドンにあるときの重さの違いを示せ．

In an electronic scale with a strain gauge load cell, the weight of an object is converted into the proportional deflection of an elastic body, from which the mass is evaluated. This type of measurement is referred to as the deflection method. This kind of scale can provide faster measurements than an electronic balance scale based on the null method since no feedback control is required, but accuracy is reduced. Heavy masses can be measured by using column type load cells. Calibration with standard weights is also necessary for this type of scale.

【Problems】

8-1）A 100 N force is applied to a stainless steel column with a sectional area of 100 mm^2 and a Young's modulus of E (Fig. 8-1). Investigate E and calculate $\Delta R_1/R_1$ of G1 (k_s=2).

8-2）Show the relationship between the ΔR of G1 and F with the parameters in Fig. 8-3, except the length of the beam.

8-3）Demonstrate that the time constant RC_d of a piezoelectric material is only dependent on the permittivity ε_{33} and the resistivity ρ of the material.

8-4）Calculate the weight difference for a 5 kg mass in Tokyo and London.

8-5）1 度の傾斜角で設置された電子天秤を使って質量が 1 kg の物体を計測したときに生じる計測誤差を計算せよ．

8-5）Calculate the measurement error when a 1 kg mass is measured by using an electronic balance with a one-degree tilt error of leveling.

第9章　材料物性値の計測

物体に力を加えると形状が変化したり，破壊したりする．力を除去した際に形状が元に戻る変形を弾性変形，元に戻らない変形を塑性変形と呼び，大きな変形の後にくびれを伴いながら発生する破壊を延性的な破壊，あまり変形を伴わずに発生する破壊を脆性的な破壊と呼ぶ．このような変形の制御と破壊の抑制は，ロボットなどのシステムにおける構造強度設計の基本であり，それらを制御するために必要となる物理量が材料強度に関わる材料物性値である．本章では，構造強度の設計と解析において重要な材料物性値として弾性変形を考慮するためのヤング率とポアソン比，塑性変形を考慮するための硬さ，破壊を考慮するための衝撃値の計測について述べる．

● 9.1　ヤング率とポアソン比の計測

静的測定法：引張り試験　丸棒試験片を軸方向に引っ張った時，試験片に負荷される

Chapter 9　Measurement of Mechanical Properties of Materials

A material is deformed and/or fractured when a force is applied to it. Deformation is called elastic or plastic deformation, depending on whether the deformation completely recovers after removal of the force. Fractures are ductile or brittle, depending on whether the fracture is accompanied by constriction after large deformations or without much deformation. The control of both deformations and fractures is an important issue in structural strength design in mechanical systems. The quantitative values pertaining to the mechanical properties of materials are essential to controlling them. In this chapter, Young's modulus, Poisson's ratio, hardness with respect to elastic and plastic deformation, and impact value with respect to fracture are described as important mechanical properties of materials for the design and analysis of the structural strength of systems such as robots.

● 9.1 Measurement of Young's modulus and Poisson's ratio

Static measurement method: Tensile test　When a round steel bar is pulled

力と試験片の伸び（変位）を前章までに詳述された計測法にて正確に測定することにより，試験片に負荷される応力（負荷された力を試験片の断面積で割った値）σ(N/m^2) とひずみ（伸びた変位を試験片の元の長さで割った値）εの関係は，例えば鉄鋼の丸棒の場合，Fig. 9-1 のように得ることができる．これを応力－ひずみ線図と呼び，JIS Z 2241「金属材料引張試験方法」に基づき図中に示す以下の値により材料強度に関わる物性値が規定される．

A 引張り強さ：試験中に加わった最大の力に対応する応力

B 降伏応力：金属材料が降伏現象を示すときの応力

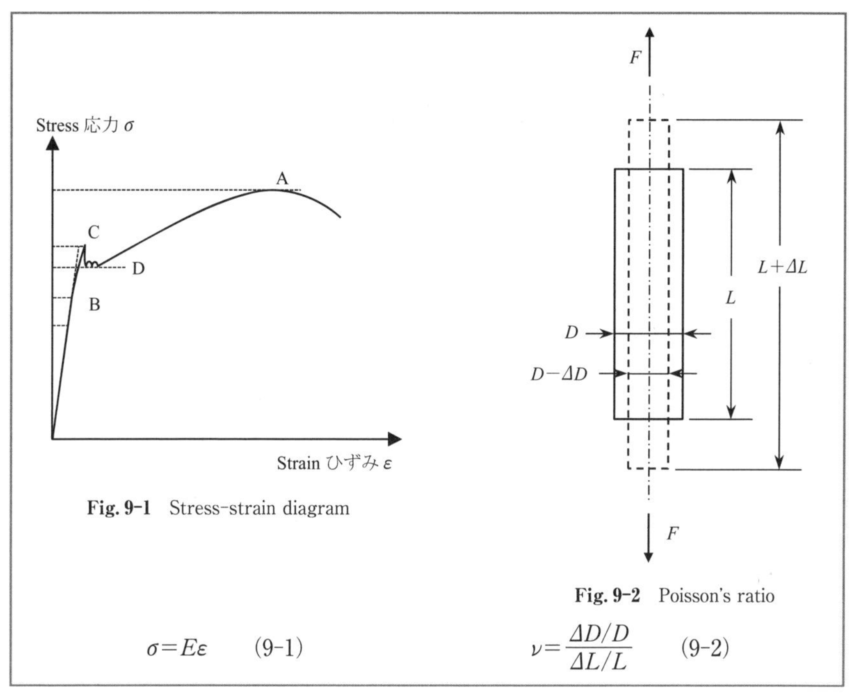

Fig. 9-1 Stress-strain diagram

Fig. 9-2 Poisson's ratio

$$\sigma = E\varepsilon \qquad (9\text{-}1)$$

$$\nu = \frac{\Delta D/D}{\Delta L/L} \qquad (9\text{-}2)$$

in the axial direction, the relationship between stress (loaded force divided by the cross-sectional area) σ (unit N/m^2) and strain (elongation divided by original length) ε is obtained as shown experimentally in Fig. 9-1. The figure is called a "stress-strain diagram"; the following four values, shown in the figure, are defined as important material physical properties by JIS Z 2241, *Metallic Materials: Tensile Testing–Method of Test at Room Temperature:*

A Tensile strength: Stress corresponding to the maximum applied force

B Yield stress: Stress at which material starts yielding

C 上降伏点：最初に力の減少が観測される瞬間の応力値

D 下降伏点：過渡的影響を無視した塑性降伏する間の応力の最小値

この図において，ひずみの小さい領域は除荷により完全に元の形状に戻る弾性変形領域であり，試験片に負荷される応力とひずみは式(9-1)に示す線形関係を示す．この時の比例定数Eがヤング率であり，力を負荷した時の弾性変形に対する抵抗を意味する材料固有の値である．また，丸棒はFig. 9-2に示すように引っ張った縦方向には延びるものの横方向には縮むこととなる．この時，式(9-2)で表される横方向のひずみ($\Delta D/D$)と縦方向のひずみ($\Delta L/L$)の比がポアソン比νであり，材料によって一定の値を示す．弾性変形域においては，ヤング率とポアソン比が重要な材料物性値となる．このような静的に大きな力が繰り返し負荷される引張り試験において求められる値の精度は，疲労や摩耗による装置としての精度に依存することとなり，その計測精度において問題はあるものの最も基本的な測定法である．

動的測定法：固有振動法，共振法　材料の固有振動数，共振周波数は，その形状とヤング率によって決定される．それゆえ，固有振動数，共振周波数の測定により材料の

C Upper yield point: Stress at which a decrease in applied stress is observed

D Lower yield point: The minimum stress required to maintain yield

In this figure, the strain is nearly proportional to the stress. This relation is expressed by Eq. (9-1) when strains are small; in other words, the region corresponds to an elastic deformation. The proportional constant E in Eq. (9-1) is the Young's modulus, which indicates the resistance of a material to elastic deformation. Further, the round bar shrinks in the transverse direction when it is pulled in the longitudinal direction, as shown in Fig. 9-2. The ratio of the strain in the transverse direction ($\Delta D/D$) to the strain in the longitudinal direction ($\Delta L/L$) is called Poisson's ratio ν; this ratio is an important material property in that, like Young's modulus, it determines elastic deformation. The accuracy of the values obtained by the tensile test depends heavily on the accuracy of the test apparatus itself, which has fatigue or wear problems owing to the repeated loading of statically large forces. Therefore, it is understood that measuring Young's modulus and Poisson's ratio by a static measurement method involves accuracy problems.

Dynamic measurement method: Natural and resonance frequency method

The natural frequency or resonant frequency of an oscillated material is determined

ヤング率を求めることが可能である．最も一般的な固有振動数の測定概略図を Fig. 9-3(a)に示す．2本の吊り線により吊られる丸棒試験片に対し相対的に大きな質量 M（単位 kg）を中央部に負荷した時の固有振動数 f（単位 s^{-1}）は式(9-3)で表される．ここで，l は2本の吊り線により支持される部分の長さ（単位 m），d は丸棒の直径（単位 m）である．よってヤング率は，固有振動数 f を式(9-4)に代入して求められる．続いて，一般的な自由共振を利用した共振法の測定概念図を Fig. 9-3(b)に示す．試験平板を2本の吊り線で保持し，駆動源から吊り線を介し固有振動を発生させ，それを他方の吊り線を介し検出する．2本の吊り線部以外は試験片に対し接触する部分

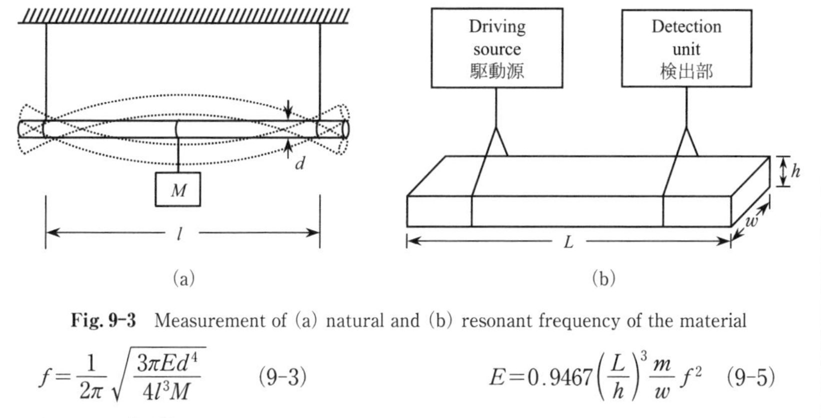

Fig. 9-3 Measurement of (a) natural and (b) resonant frequency of the material

$$f=\frac{1}{2\pi}\sqrt{\frac{3\pi Ed^4}{4l^3M}} \qquad (9\text{-}3)$$

$$E=0.9467\left(\frac{L}{h}\right)^3\frac{m}{w}f^2 \qquad (9\text{-}5)$$

$$E=\frac{16\pi}{3}\left(\frac{l}{d}\right)^3\frac{M}{d}f^2 \qquad (9\text{-}4)$$

by its shape and Young's modulus. Therefore, it is possible to determine the Young's modulus of a material by measuring the natural frequency or resonant frequency of the oscillated material. A schematic diagram to measure the natural frequency during the oscillation of a material in the form of a round rod is shown in Fig. 9-3(a). The rod is simply supported at its ends and is heavily loaded with a large mass M (unit kg) in the center. The frequency of oscillation of the rod f (unit s^{-1}) is given by Eq. (9-3), where l is the distance between the supports of two suspension lines (unit m) and d is the diameter of the rod (unit m). From these data, the Young's modulus is calculated with Eq. (9-4). A schematic diagram of the most common system used in the resonance frequency method is shown in Fig. 9-3(b). A plate held by two suspension lines generates a natural vibration through one suspension line from the driving source. The natural vibration is then detected through the other suspension line. This method has the potential for highly accurate measurements because there is nothing

がないため精度の高い測定が可能となる．この平板のヤング率Eは，測定される一次共振周波数f（単位s^{-1}）を式(9-5)に代入することにより求められる．ここでL, h, wはそれぞれ試験片の長さ，厚さ，幅（単位m）であり，mは質量（単位kg）である．

動的測定法：超音波パルス法　Fig. 9-4に超音波パルス法による測定装置の概略図を示す．約1～20 MHzの超音波パルスを試験片に伝播させ，試験片内を伝播する縦波および横波の伝播速度を計測することによりヤング率とポアソン比を式(9-6)および式(9-7)で求めることができる．ここでρは密度（単位kg/m^3），V_LとV_Sは，それぞれ縦波音速と横波音速（単位m/s）である．本手法における超音波の入出力には探触子が用いられる．試験片に変形力を加える構造を持たないため，内部摩擦などによる計測上の誤差の影響は少なく精度の高い測定が可能となる．

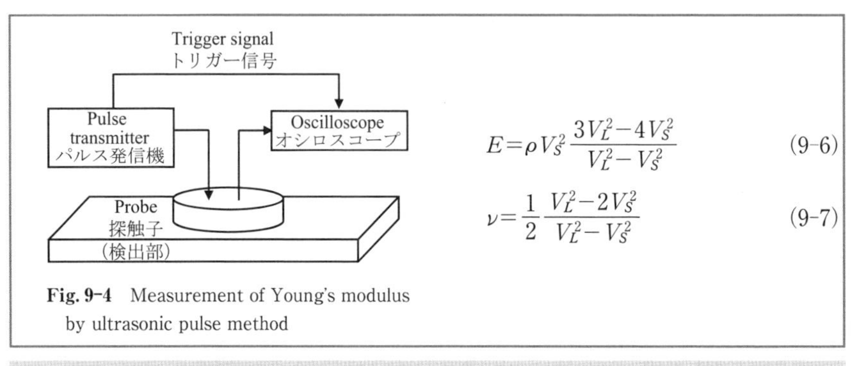

$$E=\rho V_S^2\frac{3V_L^2-4V_S^2}{V_L^2-V_S^2} \quad (9\text{-}6)$$

$$\nu=\frac{1}{2}\frac{V_L^2-2V_S^2}{V_L^2-V_S^2} \quad (9\text{-}7)$$

Fig. 9-4　Measurement of Young's modulus by ultrasonic pulse method

in contact with the plate except the two suspension lines. The Young's modulus of the plate is calculated by substituting the primary resonance frequency f (unit s^{-1}) in Eq. (9-5). Here l, h, and w are the length, thickness, and width of the plate (unit m) respectively, while m is the mass of the plate (unit kg).

Dynamic measurement method: Ultrasonic pulse method　Fig. 9-4 shows a schematic diagram for measuring the propagation velocity of longitudinal and transverse waves through the material by using the ultrasonic pulse method. By introducing an ultrasonic pulse to the test sample, the Young's modulus and Poisson's ratio are calculated using Eqs. (9-6) and (9-7) respectively; the velocity of the longitudinal and transverse waves propagating through the sample are measured. Here, ρ is the density (unit kg/m^3) and V_L and V_S are the longitudinal and transverse wave velocities (unit m/s) respectively. In this method, a probe is used for the input and output of the ultrasonic wave. Because the ultrasonic pulse does not physically

固有振動数法や共振法に代表されるこのような材料の動的挙動による測定法は，静的測定法の問題点を払しょくすることのできる精度の高い計測法である．

● 9.2　硬さの計測

鋭利な剛体を一定の力で押しあて引っ掻いた時，深い傷がつくものが軟らかい材料であり傷がつきにくいものが硬い材料と理解することができる．その意味では，硬さは塑性変形の程度を表す物理量といえる．一方，負荷した荷重を除荷するとほぼ元の形状に戻るものの応力とひずみに強い非線形性を有し，かつ塑性的な変形を与えることが困難であるゴムなどの軟質物質においては，一定の力で押しこんだ時の変形量が大きいものを軟らかい材料，変形量が小さいものを硬い材料と理解する．このような軟質物質やヤング率の非常に小さい弾性体において，硬さは弾性的変形の程度を表す物理量といえる．すなわち，対象とする材料の種類によって硬さの物理的意味は異なっており，それら材料と目的に応じた複数の硬さ測定法が存在する．

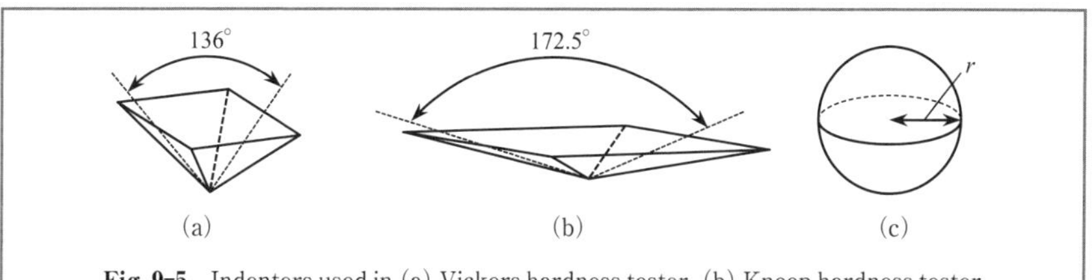

Fig. 9-5　Indenters used in (a) Vickers hardness tester, (b) Knoop hardness tester and (c) Brinell hardness tester

deform the sample, a highly accurate measurement of Young's modulus is theoretically possible.

● 9.2　Measurement of hardness

When a material is scratched by a sharp, rigid object with a constant force, it is generally understood that soft materials develop deep scratches and hard materials develop shallow scratches. In this sense, hardness is recognized as a physical quantity representing the degree of plastic deformation. On the other hand, for a soft material such as rubber, which is difficult to deform plastically, hardness is recognized as a physical quantity that represents the degree of elastic deformation. That is, the actual physical meaning of hardness depends on the type of material. There are several methods for determining hardness based on the type of material being tested and the application for which it is being considered.

ビッカース硬さ HV，ヌープ硬さ HK，ブリネル硬さ HB（JIS：R 1610，Z 2243，Z 2244，Z 2251）　これら3種類は，剛体である一定の形状を有する圧子を試験荷重で押し込み圧痕が形成される時の平均圧力によって硬さが定義される．すなわち圧子による塑性変形時の圧力（試験荷重を圧痕面積で除した値）が硬さを意味するものであり，一般に押し込み型と呼ばれる．

ビッカース硬さは，頂角 136° の正四角錘ダイヤモンド，ヌープ硬さは頂角 172.5° の四角錘（対角線長比 1：7.11）ダイヤモンド，ブリネル硬さは鋼または超硬合金の球をそれぞれ圧子としており（Fig. 9-5），それぞれの圧子により形成される圧痕形状

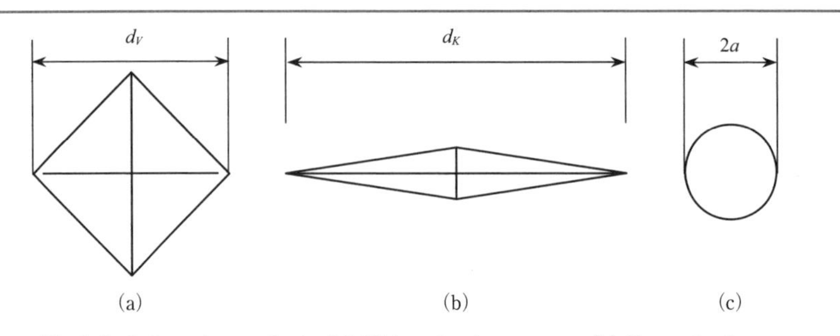

Fig. 9-6　Indentation marks in (a) Vickers hardness tester, (b) Knoop hardness tester and (c) Brinell hardness tester

$$\text{Vickers hardness: HV} = 0.102 \times \frac{2W \sin(136°/2)}{d_V{}^2} \quad (9\text{-}8)$$

$$\text{Knoop hardness: HK} = 0.102 \times \frac{W}{c d_K{}^2} \quad (9\text{-}9)$$

$$\text{Brinell hardness: HB} = 0.102 \times \frac{W}{2\pi r(r - \sqrt{r^2 - a^2})} \quad (9\text{-}10)$$

Vickers hardness HV, Knoop hardness HK, Brinell hardness HB (JIS：R 1610, Z 2243, Z 2244, Z 2251)　The hardness in these three cases is defined as the average contact pressure at an indentation caused by an indenter-applied load. The value of the hardness is indicated by the contact pressure required for plastic deformation; this pressure is calculated by the applied load divided by the area of the indentation caused by an indenter of a specific shape, as shown in Fig. 9-5. For practical reasons, each of these methods is divided into a range of scales defined by a combination of applied loads and indenter geometry. For determining Vickers hardness, Knoop hardness, and Brinell hardness, a square pyramid diamond with an angle of 136° between opposite faces, a quadrangular pyramid diamond with an angle of 172.5° (diagonal length ratio of 1：7.11), and steel or cemented carbide sphere are

(Fig. 9-6) に基づき硬さは式(9-8)～(9-10)で求められる．ここで，W は試験荷重（単位N），d_v と d_K は，それぞれビッカース圧子による圧痕の対角線長とヌープ圧子による圧痕の長い方の対角線長（単位mm），c は，ヌープ圧子の圧子定数（=0.07028），r はブリネル圧子球の半径（単位mm），a はブリネル圧子による圧痕の半径（単位mm）である．ここで，求められる値は，kgf/mm^2 の応力の次元を有しているものの，一般には，単位を記載せずに使用されることに注意が必要である．ビッカース硬さは，ブリネル硬さと比較し，2桁ほど低い荷重における μm オーダの圧痕による硬さ測定が可能である．さらにヌープ圧子はビッカース圧子と比較し，小さな押し込みに対しても大きな圧痕が形成されるため薄膜など薄い材料における測定に有効である．

ロックウェル硬さ HR（JIS：K 7202-2，Z 2245）　ビッカース硬さ，ヌープ硬さ，ブリネル硬さは，剛体を押し込んだ際に形成される圧痕の面積を測定することに

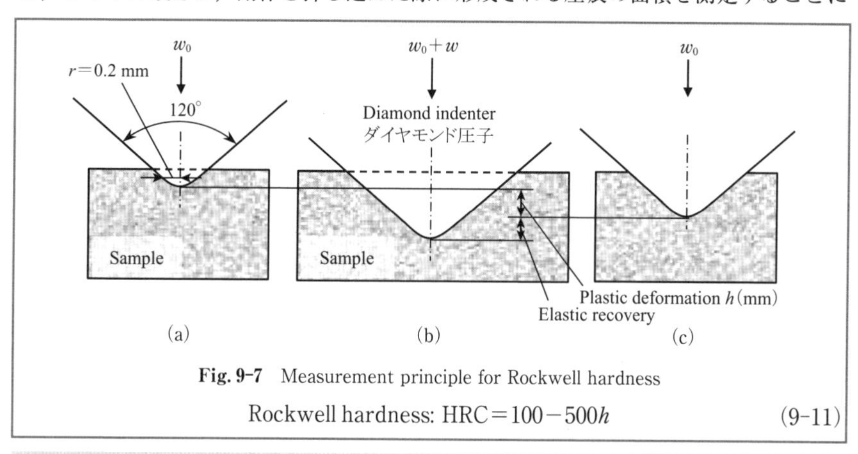

Fig. 9-7　Measurement principle for Rockwell hardness

$$\text{Rockwell hardness: HRC} = 100 - 500h \tag{9-11}$$

used as indenters respectively. The hardness value for each method is obtained using Eqs. (9-8), (9-9), and (9-10) respectively, with the shape of each indentation mark shown in Fig. 9-6. Here, W is the test load (unit N), and d_v and d_K are the diagonal lengths of the indentation caused by the Vickers indenter and the longer diagonal length of the indentation caused by the Knoop indenter (unit mm), as shown in Fig. 9-6(a) and (b); c is the constant for the Knoop indenter (=0.07028), r is the radius (unit mm) of the Brinell indenter's sphere (Fig. 9-5(c)), and a is the radius (unit mm) of the indentation mark (Fig. 9-6 (c)). It should be further noted that the hardness value is dimensionless, although the calculated values have a unit that includes a stress component (kgf/mm^2).

より硬さを求めているのに対し，ロックウェル硬さは，押し込み深さを測定することにより硬さを定義する．

ダイヤモンド円錐または鋼球の圧子を Fig. 9-7 に示すとおり測定する押し込み深さのゼロ点として初期荷重 w_0（98.07 N）を負荷した点を基準とし（Fig. 9-7(a)），さらに試験荷重を負荷し（$w+w_0$）（Fig. 9-7(b)），再び初期荷重 w_0 に戻した際の（Fig. 9-7(c)）深さの差 h により硬さを定義する．測定が簡便で測定者による誤差要因が少ないのが大きな特徴といえる．硬さの記号は HR で表され，硬質材料から相対的に軟質材料までの硬度を定義するため，A, B, C などのスケール（圧子の種類と試験荷重）を示す．例えば円錐圧子を用いて 150 kgf の試験荷重で測定を行う C スケールでは，HRC と表記し，式(9-11)でその硬度を算出する．

デュロメータ（JIS：K 6253, K 7215） ゴムやプラスティックなどの軟質材や弾性体を対象とする硬度測定法である．Fig. 9-8(a)のとおり加圧面と測定物を密着さ

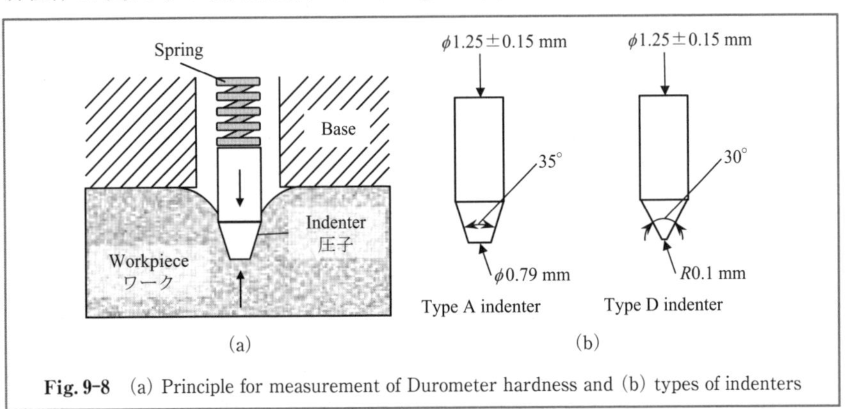

Fig. 9-8 (a) Principle for measurement of Durometer hardness and (b) types of indenters

Rockwell hardness HR (JIS: K 7202-2, Z 2245) Rockwell hardness is calculated by measuring the indentation depth due to an indenter applied load. An indenter with a diamond cone or hardened steel ball is forced into the test sample under a preliminary minor load w_0 (usually 98.07 N) (Fig. 9-7 (a)) to set a datum position. While the preliminary minor load is still being applied, an additional major load w is applied, with a resulting increase in penetration (Fig. 9-7 (b)). The removal of the additional major load allows for a partial recovery, thus reducing the depth of penetration (Fig. 9-7 (c)). The permanent increase in the depth of penetration, which results from the application and removal of the additional major load, is used to calculate the Rockwell hardness number, which is expressed by Eq. (9-11).

せ，内部にあるスプリングにより生じた加圧力により圧子が測定物に変形を与える時，測定物の反発力により平衡状態となる．この時の圧子の押し込み量により硬さが定義される．すなわち，測定物の反発力の大きさが硬さとして評価される．測定材質と目的により内部のスプリングや圧子の形状（Fig. 9-8(b)）が決定される．圧子と荷重によってタイプ A，D，E などがあり，硬さ記号はそれぞれ HAD，HDD，HED と表記される．

ショア硬さ HS（JIS：Z 2246）　圧子を押し込んだ時の抵抗や変形量で硬さを評価するのに対し，Fig. 9-9 に示すように一定の形と質量を有するハンマーを一定高さから落下させ，その跳ね上がり高さによって硬さを評価するのが反発硬さである．こ

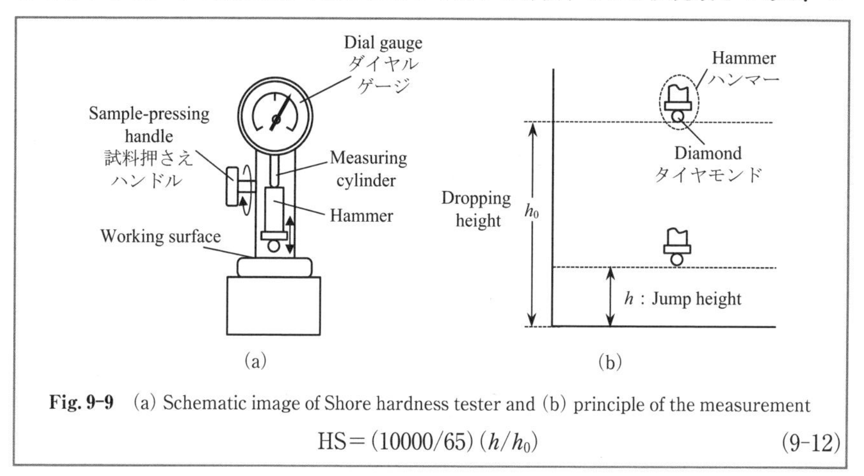

Fig. 9-9　(a) Schematic image of Shore hardness tester and (b) principle of the measurement

$$HS = (10000/65)(h/h_0) \tag{9-12}$$

Durometer hardness (JIS: K 6253, K 7215)　This is a hardness measurement method specifically intended for soft and elastic materials such as rubber and plastics. Fig. 9-8(a) shows the principle involved in determining durometer hardness. The hardness is defined by the indentation depth, which represents the repulsive force of the sample. The combination of the spring and shape of the indenter, shown in Fig. 9-8(b), is selected based on the test material and purpose of the measurement.

Shore hardness HS (JIS: Z 2246)　Fig. 9-9(a) shows a schematic diagram of the Shore hardness tester. Shore hardness is evaluated by the repulsion of the material, which is determined as a value proportional to the jump height when a hammer embedded with a diamond chip is allowed to fall on the surface of the sample from a certain height, as shown in Fig. 9-9(b). The value is calculated by Eq. (9-12).

の時の硬さをショア硬さと呼び，式(9-12)で定義される．ここで，h_0は初期高さ(m)，hは跳ね上がり高さ(m)である．精密性より携帯性などの利便性の良い硬さ試験機といえる．

● 9.3　じん性・脆性の計測

脆性破壊を引き起こす性質はもろさまたは脆性と呼ぶのに対し，脆性破壊に対する材料の抵抗の程度を粘り強さまたはじん性と呼び，それらを定量的に表す物性値がシャルピー衝撃値と破壊じん性値である．Fig. 9-10(a)に示す試験片を用い衝撃を与えて破壊するために必要なエネルギーから，材料のもろさ，粘り強さを定量的に理解す

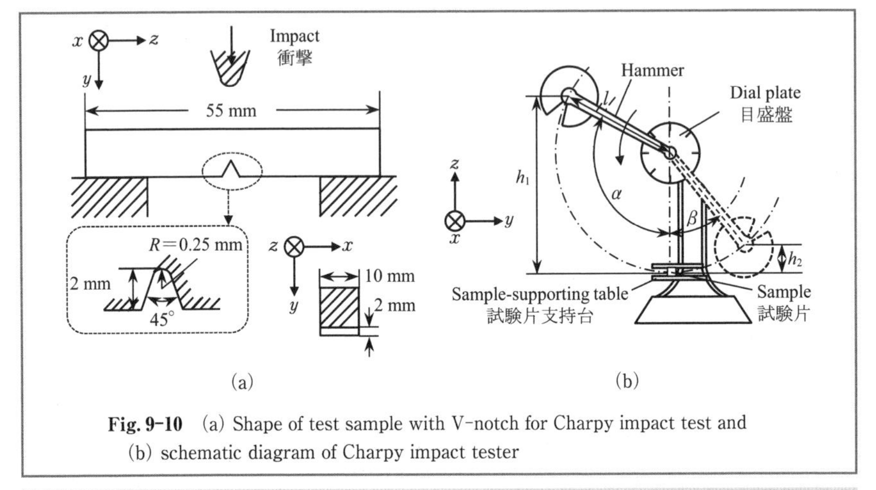

Fig. 9-10　(a) Shape of test sample with V-notch for Charpy impact test and (b) schematic diagram of Charpy impact tester

Here, h_0 and h are the initial and jumped heights (unit m) respectively. This is a portable and convenient hardness tester.

● 9.3 Measurement of toughness and brittleness

The property that causes brittle fractures is called fragility or brittleness, while the resistance of a material to a brittle fracture is referred to as tenacity or toughness. The Charpy impact value and fracture toughness are representative material physical properties with respect to fragility, tenacity and toughness of the materials. The fragility and tenacity of a material are evaluated quantitatively in terms of the energy required to break the test piece shown in Fig. 9-10(a). A schematic diagram of the Charpy impact testing machine, which conveniently measures the energy required to break the sample shown in Fig. 9-10(a) because of the impact, is shown in Fig. 9-10(b). In this test, a hammer swings down from a height of h_1 and rises to a

ることができる．衝撃により破断するために必要なエネルギーを簡便に計測するシャルピー衝撃試験機の概略を Fig. 9-10(b) に示す．ハンマーを高さ h_1 から振り下ろすと，ハンマーは切り込みを付けた試験片を破壊して高さ h_2 まで振り上がる．この時の位置エネルギーの差は，試験片を破壊するためのエネルギー E とハンマーが振り下ろされた際の空気抵抗や軸部の摩擦抵抗に起因した損失エネルギー L により式(9-13)で表されるため試験片を破壊するエネルギー E（単位 J）は式(9-14)で求められる．ここで m はハンマー重量（単位 g），g は重力加速度（単位 m/s^2），l はハンマー回転中心からハンマー重心までの距離（単位 m），L は回転する際の損失エネルギー（軸部の摩擦，空気抵抗など）（単位 J）である．なお破壊の抵抗の程度を表すシャルピー衝撃値（単位 J/cm^2）は，試験片を破壊するためのエネルギー E を切り欠き部の断面積で除したもので表される．

$$mg(h_1-h_2)=E+L \quad (9\text{-}13) \qquad E=mgl(\cos\beta-\cos\alpha)-L \quad (9\text{-}14)$$

height of h_2 after breaking the sample with a V-notch, as shown in Fig. 9-10(b). The difference in the potential energy of the system before and after breaking the sample is related to the energy required to break the specimen, which is expressed by Eq. (9-13). Further, the energy E (unit J) required to break the specimen is expressed by Eq. (9-14). Here, m is the weight of the hammer (unit g) and g is the gravity acceleration (unit m/s^2), l is the distance between the rotational center of the hammer and the center of the hammer mass (unit m), and L is the friction loss of the system (unit J). Thus, the Charpy impact value (unit J/cm^2) showing the material's resistance to fracture is evaluated in terms of the energy E (unit J) required to break the specimen, divided by the cross-sectional V-notch area of the specimen.

【演習問題】

9-1） Fig. 9-3(a)において，中央部に1 kgの重りを付けた直径10 mmを有する鉄鋼の丸棒を，30 cmの間隔で糸で吊るした時の丸棒の固有振動数が66.5 Hzであった．この鉄鋼材料のヤング率を求めよ．また，この手法によりヤング率を高精度に求めることができる理由を述べよ．

9-2） 先端が四角錐のビッカース圧子をある材料表面に1 Nで押し付けた時，圧痕の対角線長さが30 μmであった．この材料のビッカース硬度を求めよ．

9-3） 測定者による誤差要因が少ない硬さの測定法を述べよ．またその理由を述べよ．

9-4） セラミックスに代表される硬質材料とゴムに代表される軟質材料における硬さの物理的意味の違いを述べよ．

【Problems】

9-1） In Fig. 9-3 (a), a steel rod with 10 mm diameter is supported by two lines at distance of 30 cm and loaded with 1 kg of weight at the center. The frequency of oscillation of the rod is 66.5 Hz. Calculate the Young's modulus of the steel material.

9-2） The diagonal length of the indentation mark on a material surface caused by a Vickers indenter with a square pyramid diamond with a load of 1 N, is 30 μm. Calculate the Vickers hardness of the material.

9-3） Describe the hardness testing method that can provide a reproducible value despite errors introduced by the operator.

9-4） Describe the differences in the physical meanings of hardness between soft materials such as rubber and hard materials such as ceramics.

第10章　流体の計測

本章では，流体力学に基づく液体や気体などの流体の運動や諸量の把握において重要な流速と流量，ならびに圧力と真空度，さらには粘度の計測法について述べる．流速とは流体が移動する速度であり，流量とは流体が移動する量を意味し，体積流量と質量流量がある．なお，温度や湿度については第11章で説明する．

● 10.1　流速と流量の計測

流体現象を理解するためには速度場を把握する必要があるので，本章ではまず局所的な流速の計測が可能な計測法について説明する．

ピトー管　流路断面AとBにおいて，それぞれの速度をv_A，v_B，圧力をp_A，p_B，計測点の高さをy_A，y_B，流体の密度をρ，重力加速度をgとすると，ベルヌーイの

$$\frac{1}{2}\rho {v_A}^2+p_A+\rho g y_A=\frac{1}{2}\rho {v_B}^2+p_B+\rho g y_B=const. \quad (10\text{-}1)$$

Chapter 10　Measurement of Fluid

The measurement of parameters such as flow velocity, flow rate, pressure, degree of vacuum, and viscosity is very important in understanding the motions and values of liquids and gases based on fluid dynamics. Flow velocity and flow rate mean moving speed and quantity of liquid respectively. Temperature and humidity are described in Chapter 11.

● 10.1　Measurement of flow velocity and flow rate

As measurement of flow velocity field is required in order to understand the flow field, the methods to measure flow velocity are employed at local points.

Pitot tube　Bernoulli's equation is expressed by Eq. (10-1), where velocities, pressures, and heights of measurement points are shown as v_A and v_B, p_A and p_B, and

定理より式(10-1)が成り立つ．なお，流路断面 A と B が近接している場合には $y_A = y_B$ とみなすことができる．Fig. 10-1 には，ピトー管と静圧管を組み合わせた代表的な二重管を示す．中心部の管の先端を流れに向けて全圧 p_t を計測するとともに，管の側面に孔を設けて静圧 p_s を計測すれば，流速 v，全圧 p_t，静圧 p_s にはベルヌーイの式(10-1)から式(10-2)の関係が成り立つので，流速 v は全圧 p_t，静圧 p_s の差から式(10-3)で求めることができる．

熱線流速計　乱流の計測などの気体の動的な流速の計測には，流速により熱した線が冷却されて温度が低下し，この温度低下により熱線の抵抗が変化する原理を用いた熱線流速計が用いられる．要するに，流速を熱の移動量，すなわち熱線の電気抵抗の変化として計測する．Fig. 10-2 に示すような直径数 μm から数十 μm の白金線やタング

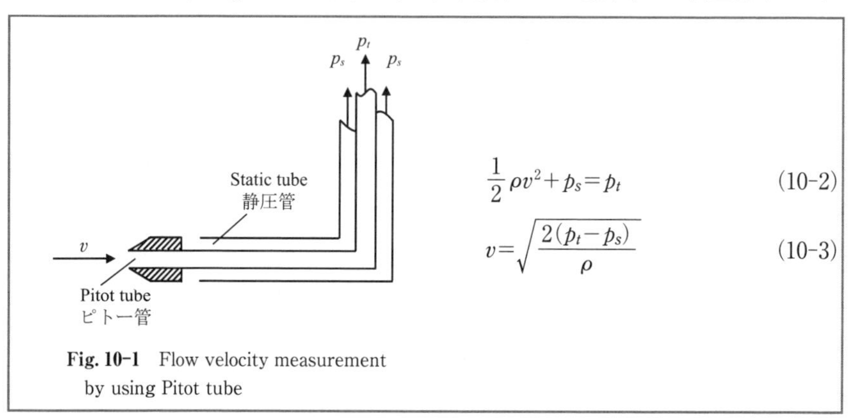

$$\frac{1}{2}\rho v^2 + p_s = p_t \qquad (10\text{-}2)$$

$$v = \sqrt{\frac{2(p_t - p_s)}{\rho}} \qquad (10\text{-}3)$$

Fig. 10-1　Flow velocity measurement by using Pitot tube

y_A and y_B respectively at the cross section of A and B. Here, ρ and g are density of liquid and acceleration of gravity. If A and B are very close, it is assumed that $y_A = y_B$. Fig. 10-1 shows the schematic of a duplex tube made up of a Pitot tube and static pressure tube. The flow velocity v is obtained from the pressure difference between p_t and p_s in Eq. (10-3), as there is a relationship between flow velocity v, total pressure p_t, and static pressure p_s, as shown in Eq. (10-2) from Bernoulli's equation, by measuring total pressure p_t and static pressure p_s.

Hot-wire anemometer　In order to measure flow velocity due to turbulence, a hot-wire anemometer is used. As the hot wire is cooled by the flow velocity and the resistance of the hot wire changes when cooling down, the flow velocity can be evaluated by measuring the resistance of the wire. The wire whose diameter is several mm or several dozen μm and be made of metal such as platinum, tungsten, or

ステン線，ニッケル線などが用いられる．熱線周囲の流体の速度が v，温度が T_s のとき，長さ l，直径 d，温度 T_w の熱線から周囲の流体に伝達される熱量は，式(10-4)で表される．ここで，k は熱伝導率，ρ は流体の密度，c_p は流体の定圧比熱である．なお，熱線の抵抗が R_w，熱線に流れる電流が i のとき，熱線から与えられる熱量は式(10-5)であるので，平衡状態のとき式(10-4)と(10-5)から，流速 v は式(10-6)で求めることができる．ここで，a，b は式(10-4)を変形した際の定数である．なお，熱線の温度 T_w は，基準温度 T_0 における熱線の抵抗 R_0 と熱線の抵抗温度係数 α から求めることができる．なお，計測時に熱線を平衡状態に保つために，温度を一定にする定温度法と，電流を一定にする定電流法があるが，定温度法が主流である．

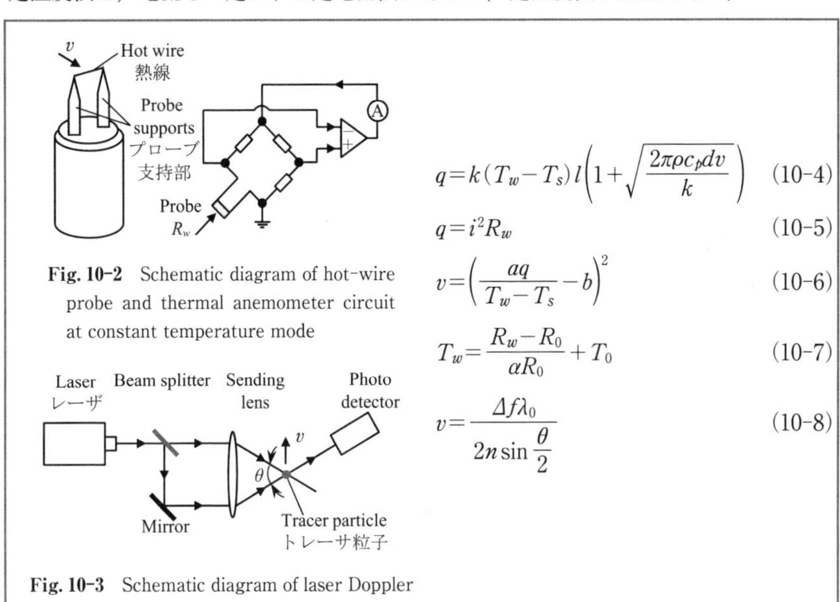

Fig. 10-2　Schematic diagram of hot-wire probe and thermal anemometer circuit at constant temperature mode

Fig. 10-3　Schematic diagram of laser Doppler

$$q=k(T_w-T_s)l\left(1+\sqrt{\frac{2\pi\rho c_p dv}{k}}\right) \qquad (10\text{-}4)$$

$$q=i^2R_w \qquad (10\text{-}5)$$

$$v=\left(\frac{aq}{T_w-T_s}-b\right)^2 \qquad (10\text{-}6)$$

$$T_w=\frac{R_w-R_0}{\alpha R_0}+T_0 \qquad (10\text{-}7)$$

$$v=\frac{\Delta f\lambda_0}{2n\sin\dfrac{\theta}{2}} \qquad (10\text{-}8)$$

nickel, as shown in Fig. 10-2. The heat transfer from a wire of length l, diameter d, and temperature T_w is expressed by Eq. (10-4) at the flow velocity v and temperature T_s of surrounding fluid. Here, k, ρ, and c_p are heat conductivity, density, and isobaric specific heat of fluid respectively. As heat transfer from the wire of resistance R_w and current i is expressed by Eq. (10-5), the flow velocity is obtained from Eq. (10-6) at an equilibrium condition. Here, a and b are constants obtained by changing Eq. (10-4). The temperature of the wire T_w can be obtained from the resistor temperature coefficient α and the resistance R_0 at base temperature T_0 of the wire. In order to keep the wire at equilibrium condition, there are constant temperature methods and constant current methods, the constant temperature method is more commonly used.

レーザドップラー流速計　ピトー管や熱線流速計による流速の計測ではプローブを流れ場に挿入する必要があるが，7.1節で説明したドップラー効果を用いたレーザドップラー流速計では非接触で流速を計測することができる．なおレーザドップラー流速計では，ドップラー信号を得るために，トレーサ粒子を流れ場に入れる必要がある．トレーサ粒子の速度 v は，ドップラー周波数シフト Δf と，Fig. 10-3に示すビームスプリッタで2つのビームに分けたビーム角 θ から式(10-8)で求めることができる．ここで，n は流体の屈折率，λ_0 は使用する連続レーザの波長である．

画像処理式流速計測法　Fig. 10-4に示すようなパルスレーザと高速度ビデオカメラの撮影システムにより時刻 t_1 と t_2 で撮影した画像を用いた画像処理により流速を計測できる．2枚の画像内の特定領域内の画像の相関から流速を求める画像相関法と，粒子追跡法がある．画像相関法では，式(10-9)で示すように時刻 t_1 の画像の特定領

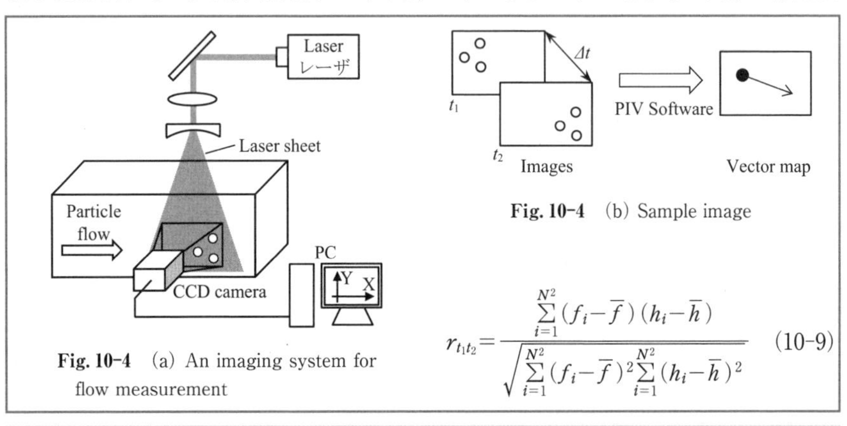

Fig. 10-4　(a) An imaging system for flow measurement

Fig. 10-4　(b) Sample image

$$r_{t_1t_2}=\frac{\sum_{i=1}^{N^2}(f_i-\overline{f})(h_i-\overline{h})}{\sqrt{\sum_{i=1}^{N^2}(f_i-\overline{f})^2\sum_{i=1}^{N^2}(h_i-\overline{h})^2}} \qquad (10\text{-}9)$$

Laser Doppler velocimeter　Although a probe can be inserted in the flow field at the Pitot tube and hot wire anemometer, the flow velocity can also be measured by the laser Doppler method, described in Section 7.1, without inserting the probe. In order to obtain a signal from the flow field, tracer particles are required. The velocity of a tracer particle can be obtained by Eq. (10-8) from Doppler shift Δf and beam angle θ, as shown in Fig. 10-3. Here, n and λ_0 are the refraction index of liquid and the wavelength of the laser used respectively.

Flow velocity measurement by using image analysis　A flow velocity field can be measured from two pictures at t_1 and t_2 by image analysis using a pulse laser and high-speed video camera, as shown in Fig. 10-4. There are two such methods, the

域の輝度 f_i と時刻 t_2 の輝度 h_i について検索領域 $N\times N$ で相関係数 $r_{t_1t_2}$ を求め，$r_{t_1t_2}$ が最大になる時刻 t_1 の画像の位置と時刻 t_2 の画像の位置から速度ベクトルを求める．粒子追跡法では，時刻 t_1 の画像上の粒子と時刻 t_2 の画像上の粒子の位置を特定して，それらの位置から速度ベクトルを求める．

次に，管路や流路を流れる流量を計測する方法について述べる．

差圧式流量計　Fig. 10-5 に示すように管にオリフィス，ノズル，ベンチュリ管などの絞り機構を設置して流路断面 A と B の圧力差 p_A-p_B を計測すれば，流量を計測できる．管断面の流速を一定と仮定し，A，B の断面積をそれぞれ S_A，S_B とすると式(10-10)が成り立ち，ベルヌーイの定理の式(10-1)より，流量 Q は差圧と絞り面積

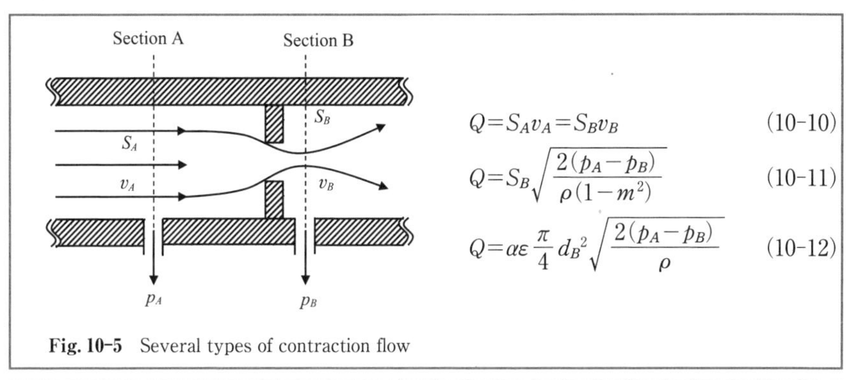

$$Q=S_Av_A=S_Bv_B \tag{10-10}$$

$$Q=S_B\sqrt{\frac{2(p_A-p_B)}{\rho(1-m^2)}} \tag{10-11}$$

$$Q=\alpha\varepsilon\frac{\pi}{4}d_B{}^2\sqrt{\frac{2(p_A-p_B)}{\rho}} \tag{10-12}$$

Fig. 10-5　Several types of contraction flow

digital image correlation method and particle image velocimetry (PIV). In the digital image correlation method, the correlation $r_{t_1t_2}$ between brightness f_i in a certain region at t_1 and h_i at t_2 within the searching area $N\times N$ is examined with Eq. (10-9), and the velocity vector is obtained when $r_{t_1t_2}$ is at its maximum. With PIV, the position of the tracer particle on the image at t_1 is defined at the position on the image at t_2, after which the velocity vector can be obtained.

Next, measurement methods of flow rate through a pipe or channel are explained.

Differential flow meter　Flow rate can be evaluated by measuring pressure difference p_A-p_B at cross sections A and B of contraction flow through an orifice, nozzle, or Venturi tube. Fig. 10-5 illustrates the contraction flow though the orifice. When uniform flow velocity at all cross sections is assumed, the flow rate Q is

比 m から式(10-11)により求まる．実際には，絞り機構で縮流を生じるため流量係数 α を考慮する必要があり，圧縮性流体である気体の計測では膨張補正係数 ε を用いて補正する必要があるので，流量は式(10-12)を用いて求める．

面積式流量計　差圧式流量計では面積比を一定にして差圧から流量を求めるのに対して，面積を変えて差圧一定として流量を求める面積式流量計があり，代表的なものはロータメータである．ロータメータは，Fig. 10-6 に示すように，テーパ管とフロートで構成され，流体は下から上に流れる．フロートの体積を V_F，断面積を S_F，密度を ρ_F とすると，フロートに作用する重力と差圧による浮力から式(10-13)が成り立

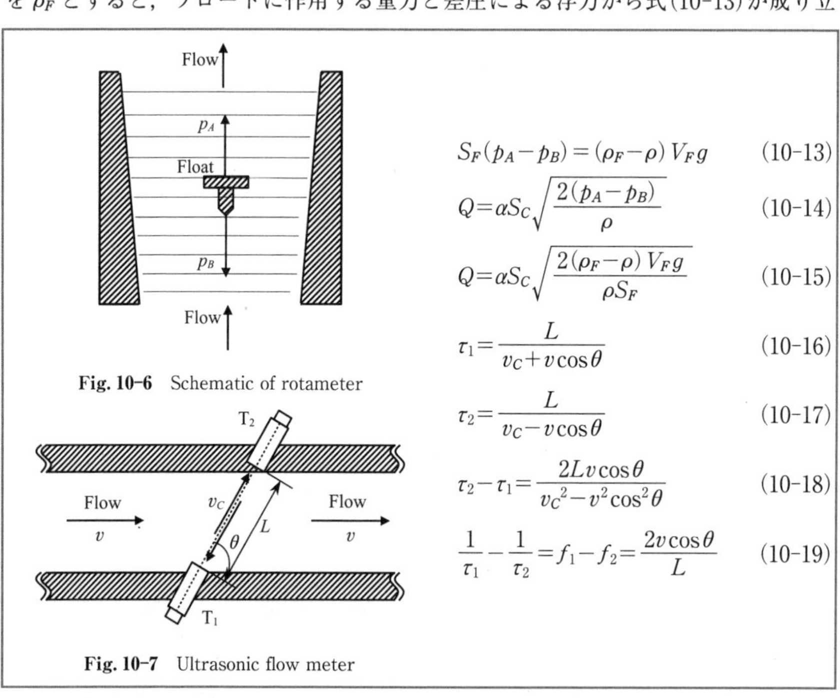

Fig. 10-6　Schematic of rotameter

Fig. 10-7　Ultrasonic flow meter

$$S_F(p_A - p_B) = (\rho_F - \rho) V_F g \quad (10\text{-}13)$$

$$Q = \alpha S_C \sqrt{\frac{2(p_A - p_B)}{\rho}} \quad (10\text{-}14)$$

$$Q = \alpha S_C \sqrt{\frac{2(\rho_F - \rho) V_F g}{\rho S_F}} \quad (10\text{-}15)$$

$$\tau_1 = \frac{L}{v_C + v\cos\theta} \quad (10\text{-}16)$$

$$\tau_2 = \frac{L}{v_C - v\cos\theta} \quad (10\text{-}17)$$

$$\tau_2 - \tau_1 = \frac{2Lv\cos\theta}{v_C{}^2 - v^2\cos^2\theta} \quad (10\text{-}18)$$

$$\frac{1}{\tau_1} - \frac{1}{\tau_2} = f_1 - f_2 = \frac{2v\cos\theta}{L} \quad (10\text{-}19)$$

calculated by Eq. (10-11) from Bernoulli's equation (10-1), as the result of Eq. (10-10) is assumed. Here, S_A, S_B, and m are cross section areas at A and B and the contraction flow rate respectively. For the actual measurement, the flow rate is obtained by Eq. (10-12), as flow coefficient α should be considered because of squeezing flow, and the expansion factor ε due to compressive flow is also relevant.

Areal flow meter　With a differential flow meter, flow rate is measured from pressure differences in a constant flow area, the flow rate can be measured by

つ．フロートと管の隙間の面積をS_Cとすると，式(10-12)と同様に式(10-14)が得られ，式(10-13)を代入すると，式(10-15)により流量が計測できることが分かる．流量係数αは，フロート形状や流体の種類によって異なるので，ロータメータには各流体の専用の目盛りで記されている．

超音波流量計 音波が媒質を伝播するとき，伝播速度が媒質の移動速度に重畳することを利用して流速を計測することができる．Fig. 10-7 に示すように，流れに対して超音波の発信器T_1と受信器T_2を設置すると，T_1からT_2への到達時間τ_1とT_2からT_1への到達時間τ_2は，それぞれ式(10-16)，(10-17)となる．ここでv_Cは音速である．式(10-18)から流速vが求まるように思えるが，一般に計測対象のvに比べてv_Cが大きく温度などによるv_Cの変動の影響が大である．T_1からT_2に到達した信号をトリガ信号としてT_1からT_2に信号を送信してτ_1の周波数f_1を求め，同様にτ_2のf_2を求めて周波数の差から，速度を求めるシングアラウンド方式が用いられる．

電磁流量計 磁場の中を導体が移動すると導体に起電力が生じる電磁誘導の法則を利

changing the area at a constant pressure with an areal flow meter that consists of a tapered tube and float, as shown in Fig. 10-6. In Fig. 10-6, the flow direction is upward. The flow rate is obtained by Eq. (10-15), as Eq. (10-13) is established by the gravity and buoyancy of the float and Eq. (10-14) is obtained as in Eq. (10-12). Here, V_F, S_F, S_C and ρ_F are the volume of float, the sectional area of the float, the area between tube and float, and the density of the float respectively. The scale is plotted on the meter, as the flow coefficient α depends on the liquid and the shape of the float.

Ultrasonic flow meter Flow rate can be measured using propagating speed, as the speed is superimposed on the speed of the medium. Fig. 10-7 illustrates an ultrasonic meter with transmitter T_1 and receiver T_2. Here, τ_1 and τ_2 are propagating times from T_1 to T_2 and from T_2 to T_1 respectively, and are expressed by Eqs. (10-16) and (10-17). The sound velocity is v_C. The velocity v can be obtained by Eq. (10-18), but v_C is too large to compare with v and is affected by temperature. Thus, the "sing-around" method is used by measuring frequency f_1 of t_1 and f_2 of t_2, using the signal detected at T_2 for the trigger of a signal from T_2.

Electromagnetic flow meter Flow rate is also measured by electromagnetic

用して，流速を計測することができる（Fig. 10-8）．流量 Q は，式(10-21)から定義される管の平均流速 $\bar{v}$ と管の直径 d から式(10-20)から求めることができる．ここで r は管の中心からの距離である．磁束密度 B を流れる導体には $\bar{v}$ に応じて電位差 E が生じるため式(10-22)が成り立つので，式(10-20)に式(10-22)を代入すると，流量 Q が磁束密度 B を付与した場合の電位差 E から計測できることが分かる．

● 10.2　圧力と真空度の計測

天気予報の気圧に代表されるように，流体の状態の把握においては，圧力や真空度の計測が重要である．計測対象の圧力・真空度の範囲，センサの大きさ，簡便さ，測定精度などから種々の圧力計や真空計が用いられている．

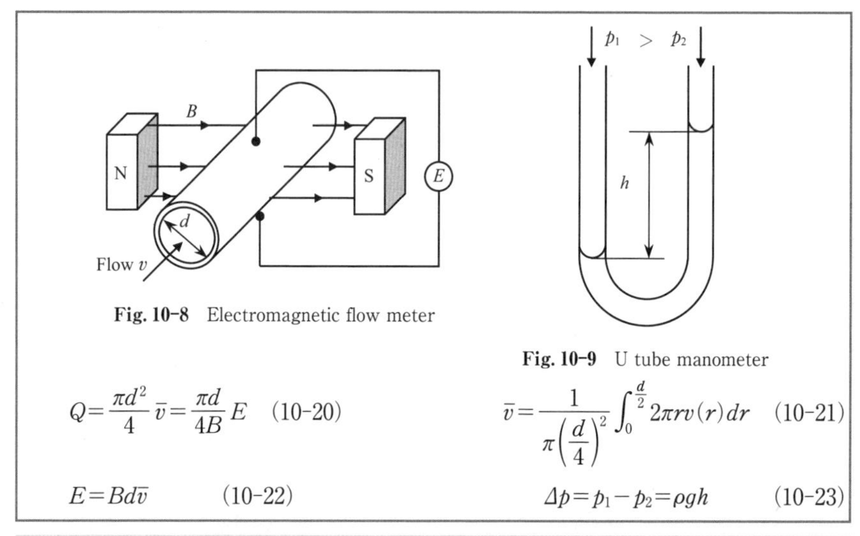

Fig. 10-8　Electromagnetic flow meter

Fig. 10-9　U tube manometer

$$Q=\frac{\pi d^2}{4}\bar{v}=\frac{\pi d}{4B}E \quad (10\text{-}20)$$

$$\bar{v}=\frac{1}{\pi\left(\frac{d}{4}\right)^2}\int_0^{\frac{d}{2}}2\pi r v(r)\,dr \quad (10\text{-}21)$$

$$E=Bd\bar{v} \quad (10\text{-}22)$$

$$\Delta p=p_1-p_2=\rho gh \quad (10\text{-}23)$$

induction, as electromotive force is produced when a conductor moves through a magnetic field. The flow rate Q is obtained from the mean flow velocity $\bar{v}$ defined by Eq. (10-21) and the mean diameter of tube d (Fig. 10-8). Here, r is the distance from the tube center. The flow rate can be measured from the potential difference E at applied magnetic flux density B, as E is proportional to B and $\bar{v}$, as shown by Eq. (10-22).

● 10.2　Measurement of pressure and vacuum

As represented by atmospheric pressure in weather forecasts, pressure and degree of vacuum are very important for understanding the state of a liquid. Several methods are used to measure the range of pressure and depend on the degree of vacuum, sensor size, convenience, and accuracy.

液柱式圧力計　流体の圧力差はマノメータと呼ばれる液柱の液面の高低差から計測することができる．Fig. 10-9 には，代表的な U 字管型を示す．U 字管型マノメータでは，式(10-23)に基づき液面間の高低差 h から圧力 p_1 と p_2 の圧力差 $\Delta p = p_1 - p_2$ を求めることができる．U 字管型の他には，一方の液中の断面積を極力大きくするために液槽に連結して，1 つの管の高さだけを計測して圧力を計測する単管型や，小さな圧力差を精度良く計測できるように単管を斜めにした傾斜管などがある．

弾性変形式圧力計　弾性変形を利用した圧力計には，Fig. 10-10 に示すようなブルドン管圧力計，Fig. 10-11 に示すようなピエゾ抵抗式圧力計，Fig. 10-12 に示すようなダイアフラム型圧力計などがある．ブルドン管圧力計は，断面が楕円形状をした中空曲管が圧力源に連結し，もう一方の端がラックとピニオン機構を介して指針に連結されており，目盛り板により圧力を読み取る．ピエゾ抵抗式圧力計は，圧力変化に対し

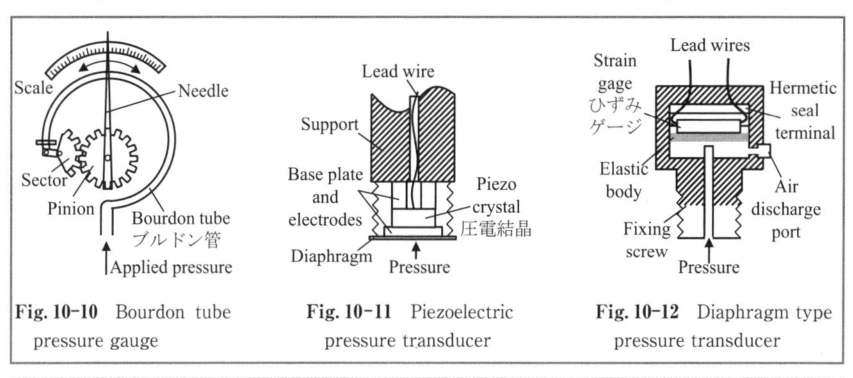

Fig. 10-10　Bourdon tube pressure gauge

Fig. 10-11　Piezoelectric pressure transducer

Fig. 10-12　Diaphragm type pressure transducer

Manometer　The pressure difference of liquid can be measured by the difference in the height of a liquid column by a manometer. Fig. 10-9 illustrates a U tube manometer. The pressure difference $\Delta p = p_1 - p_2$ can be measured from the difference of height h of liquid levels by using Eq. (10-23). A single tube manometer can measure pressure difference with a single tube, as the other end is connected to a large chamber to increase cross section area. An inclined tube manometer in which the single tube is inclined is used to measure pressure difference more precisely.

Elastic deformation type transducer　Pressure gauges using elastic deformation are illustrated in Figs. 10-10, 10-11, and 10-12. Fig. 10-10 shows a Bourdon tube pressure gauge. A curved metal tube with an elliptical cross section, which is called a Bourdon tube, is connected to a pressure source and the other end is

て電気抵抗が変化することを利用した圧力計である．またピエゾ電気効果（圧電効果）を用いて衝撃圧力を計測する圧力計もある．ダイアフラム型圧力計では，圧力源に連結されたダイアフラムにひずみゲージが貼られ，ダイアフラムの変形から圧力を計測する．また半導体のピエゾ抵抗効果を用いてダイアフラムの変形を検出して圧力を計測するセンサもある．

低真空度の計測には，前述の液柱式圧力計や弾性変形式圧力計を使用できる．ただし，液柱式圧力計を用いる場合は，低圧での蒸発を極力回避するために，蒸気圧の低い液体を使用する必要がある．100 Pa 以下の低圧では，マクラウド真空計，熱伝導真空計（ピラニ真空計），電離真空計などが用いられる（Fig. 10-13～Fig. 10-15）．

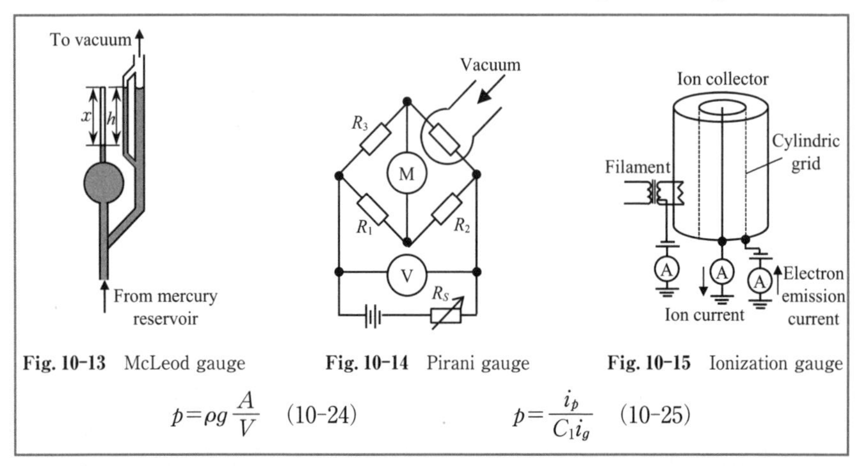

Fig. 10-13　McLeod gauge　　**Fig. 10-14**　Pirani gauge　　**Fig. 10-15**　Ionization gauge

$$p=\rho g\frac{A}{V} \quad (10\text{-}24)$$

$$p=\frac{i_p}{C_1 i_g} \quad (10\text{-}25)$$

connected to a needle through a sector and pinion. The pressure can be obtained from the needle and a scale. Fig. 10-11 shows a schematic of a piezoelectric pressure transducer, which use the piezo effect. A pressure gauge to detect dynamic impact using the piezo effect also exists. Fig. 10-12 shows a diaphragm type pressure transducer. A strain gauge is put on the diaphragm to detect pressure through the elastic deformation of diaphragm. A diaphragm with a semiconductor on the diaphragm is also commercially available.

The degree of vacuum can be measured by a manometer or an elastic deformation type sensor in low-vacuum regions. With a manometer, a liquid must have a vapor pressure low enough to avoid vaporization in low pressure conditions. At pressure conditions below 100 Pa, McLeod gauges, Pirani gauges, and ionization gauges, which are shown in Figs. 10-13 through 10-15, are used.

マクラウド真空計 閉管型の液柱式圧力計ともいえるもので，Fig. 10-13 に示す h と x が等しくなるように測定すれば，式(10-24)から圧力を求めることができる．ここで，ρ は水銀の密度，A は細管の断面積，V は細管と球状部の体積である．

ピラニ真空計 真空度が低い気体分子が衝突する回数が多い場合は加熱されたフィラメントの熱エネルギーがフィラメントに衝突する分子により持ち去られて冷却されるが，真空度が高まり気体分子の衝突回数が減少するとフィラメントの温度が上昇する原理を用いて，フィラメントの抵抗から真空度を計測する．ピラニ真空計を用いて真空度を計測する場合には，あらかじめ実験的に校正曲線を求める必要がある．

電離真空計 気体分子を電離した際に，加熱された陰極に生じるグリッド電流 i_g とプレート電流 i_p の比が圧力に依存することを用いて式(10-25)により圧力を電気的に測定する．なお，C_1 は定数である．

McLeod gauge This is a kind of closed manometer in which the degree of vacuum can be measured at $h=x$ (Fig. 10-13) by Eq. (10-24). Here, ρ, A, and V are the density of the mercury of the liquid, the cross sectional area of the capillary, and the volume of the spherical body and capillary respectively.

Pirani gauge The degree of vacuum is measured by using the relation between the number of attacks of gas molecules and the temperature of a filament. When the number of attacks of gas molecules grows larger, the temperature of the filament is decreased by removing gas molecule attacks, while the temperature is increased when the number of attacks is smaller. Thus, the degree of vacuum can be measured by the resistance of the filament that changes with temperature. With a Pirani gauge, calibration is required before testing.

Ionization gauge The degree of vacuum is measured by the relation between the ratio of grid current i_g and plate current i_p, using Eq. (10-25). Here, C_1 is constant.

● 10.3　粘度の計測

流れ場における物体に作用する力を把握するには，流体の粘度を知る必要がある．主な粘度の計測方法には Fig. 10-16 に示すように以下の2つがある．なお，粘度は温度により大きく変化する液体が多いので温度を正確に測定しておく必要がある．

毛細管粘度計　流体の粘度が高いほど流れにくい性質を利用し，半径 r，長さ L の細管を用いて，出口と入口の圧力差 Δp，体積 V の流体が流れるに要する時間 t から式(10-26)を用いて粘度 η を求めることができる．

回転粘度計　円筒に入れた流体に回転体を入れて回転させたとき，粘度が高いほど回転体にトルクがかかる性質を利用して粘度を計測する方法である．回転体の長さ L，回転体の外径 r_1，円筒の内径 r_2，回転体の角速度 ω，回転体のねじれ角 θ，回転体の

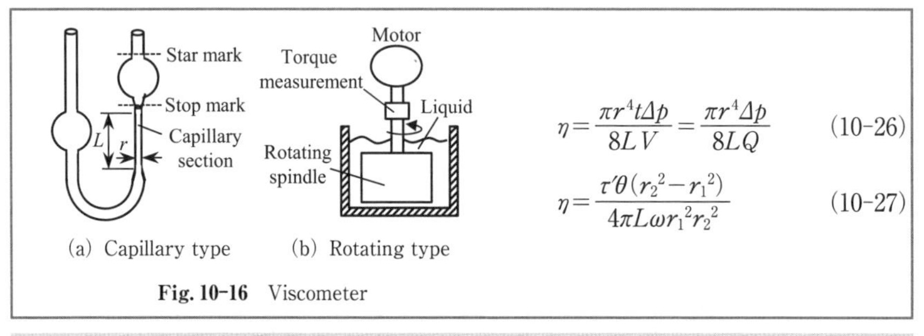

$$\eta=\frac{\pi r^4 t\Delta p}{8LV}=\frac{\pi r^4\Delta p}{8LQ} \tag{10-26}$$

$$\eta=\frac{\tau'\theta(r_2{}^2-r_1{}^2)}{4\pi L\omega r_1{}^2 r_2{}^2} \tag{10-27}$$

(a) Capillary type　　(b) Rotating type

Fig. 10-16　Viscometer

● 10.3　Measurement of viscosity

In order to estimate the applied force of a body in liquid, viscosity is an important parameter. Two typical methods, capillary and rotating, are used and shown in Fig. 10-16. Note that any temperatures measured should be written down, as viscosity changes with temperature.

Capillary type viscometer　Viscosity is measured using liquid flow. The viscosity η is obtained from the pressure difference Δp, liquid of volume V, and time to flow t with a capillary whose diameter is r and length is L, using Eq. (10-26).

Rotating type viscometer　Viscosity can also be measured by the torque of a rotating body in liquid. The viscosity η can be calculated from restoring moment τ', angular velocity ω, and twisted angle θ of a rotating body with length L and outer diameter r_1 in a cylindrical container of inner diameter r_2, by using Eq. (10-27).

ねじれ角に対する復元モーメントτ'から式(10-27)により求めることができる．

【演習問題】

10-1) ピトー管で水の流れを計測したときの全圧と静圧の圧力差が以下のとき，それぞれの速度を求めよ．

(1) 115 kPa (2) 0.416 MPa (3) 6.12 MPa

10-2) U字管型マノメータで圧力を計測したとき以下のとおりであった．それぞれの圧力を求めよ．

(1) 53 mm（水柱） (2) 23 mm（水銀柱）

10-3) 以下の圧力をPaに換算せよ．

(1) 12 mmHg (2) 1.25 bar (3) 15 Torr

10-4) 以下の動粘度を求めよ．

(1) 空気（粘度 1.80×10^{-5} Pa s，密度 1.205 kg/m^3）

(2) 水（粘度 1.002×10^{-3} Pa s，密度 9.982×10^2 kg/m^3）

【Problems】

10-1) When water flow is measured using a Pitot tube, the pressure differences of total pressure and static pressure are as follows; obtain each velocity.

(1) 115 kPa (2) 0.416 MPa (3) 6.12 MPa

10-2) When pressure is measured using a U tube manometer, the difference in height of the liquid levels are as follows; obtain each pressure.

(1) 53 mm (Water) (2) 23 mm (Mercury)

10-3) Convert the following pressures to pressure in Pa.

(1) 12 mmHg (2) 1.25 bar (3) 15 Torr

10-4) Calculate the kinetic viscosity.

(1) Air (Viscosity 1.80×10^{-5} Pa s, Density 1.205 kg/m^3)

(2) Water (Viscosity 1.002×10^{-3} Pa s, Density 9.982×10^2 kg/m^3)

第 11 章　温度と湿度の計測

温度と湿度は工業現場において重要なパラメータである．測定に用いられるセンサの小型化により，その適用範囲も急速に拡大しており，我々の身近でも，家電製品，パーソナルコンピュータやスマートフォンなどの電子デバイス内で多用されている．本章では，まず，温度および湿度の定義について触れた後，それぞれを測定するために用いられるセンサの原理および構造を紹介する．

● 11.1　温度と湿度の関係

温度は物質の熱振動をもとに規定されるパラメータである．温度のスケールとしては，セルシウス温度（単位 ℃）と，SI 単位である絶対温度（単位 K）が用いられている．セルシウス温度は 1 気圧下で水の凝固点を 0℃，沸点を 100℃として定義されており，その値は相対値である．その一方で，絶対温度では水の三重点温度 T_0

Chapter 11　Measurement of Temperature and Humidity

Temperature and humidity are important variables in modern industry. Due to the miniaturization of sensors used for measuring these variables, their application is rapidly growing. These sensors are often employed in consumer electronics and electronic devices such as personal computers or mobile phones. In this chapter, definitions of temperature and humidity are explained, after which the measurement principles and structures of selected sensors for measuring of these variables are explained.

● 11.1　Relationship between temperature and humidity

Temperature is a variable determined by the thermal motion of atoms and molecules. As a scale for expressing the temperature, the Celsius scale (unit ℃) and the Kelvin scale (unit K which is an SI unit) are often used. The Celsius scale is a relative scale based on the freezing (0℃) and boiling (100℃) points of water at the

（=273.16 K=0.01℃）を 273.16 で割ったものを 1 K として定義しており，その値は絶対値である．

湿度は気体中に含まれる水蒸気量を表すパラメータで，その指標として絶対湿度と相対湿度が用いられる．絶対湿度 d_v は，単位体積（1 m^3）の気体中に含まれる水蒸気の質量（単位 g/m^3）で定義される．水蒸気の質量 m_v，および空気の体積 V をもとに，d_v は式(11-1)によって算出できる．相対湿度 H も気体中に含まれる水蒸気量を示すが，その定義は飽和水蒸気圧 p_s と大気中の水蒸気圧 p との比とされており，式(11-2)で与えられる．単位としては%RH が用いられる．いま，水上の飽和水蒸気圧 p_s は JIS の定義より式(11-3)で示される．ここで T_a は空気の絶対温度であり，それに対応するセルシウス温度を t_a とする．普遍気体定数を R（=8314.46 J K^{-1} mol^{-1}），水蒸気のモル質量を M_v（=18.015 g/mol）として，大気中の水蒸気圧 p は式(11-4)で与えられる．また，標準気圧 P_0（=101325 Pa）を用いて，絶対湿度は式(11-5)で与

$$d_v=\frac{m_v}{V} \quad (11\text{-}1) \qquad H=\frac{p}{p_s}\times 100 \quad (11\text{-}2) \qquad p=\frac{m_v}{V}\cdot\frac{RT_a}{M_v} \quad (11\text{-}4)$$

$$\ln p_s=-6096.9385T_a^{-1}+21.2409642-2.711193\times 10^{-2}T_a+1.673952\times 10^{-5}T_a^2+2.433502\ln T_a \quad (11\text{-}3)$$

$$d_v=p\frac{M_v}{RT_a}=\frac{18.015p}{0.0831446(273.15+t_a)}=\frac{803.7}{1+0.00366t_a}\cdot\frac{p}{P_0} \quad (11\text{-}5)$$

pressure of 1 atm. Meanwhile, the Kelvin scale is an absolute scale defined as the fraction 1/273.16 of the thermodynamic temperature of the triple point of water T_0 (exactly 0.01℃ and 273.16 K).

Humidity is a variable determined by the amount of water vapor in the air. There are two kinds of humidity scales: absolute humidity and relative humidity. Absolute humidity d_v corresponds to the amount of water in the air regardless of temperature; it is expressed in grams per cubic meter (g/m^3). Eq. (11-1) gives d_v, with the mass of the water vapor m_v and the volume V of the air. Relative humidity H also corresponds to the amount of water in the air but is relative to the maximum for a given temperature (unit %RH). The equilibrium vapor pressure of water p_s (unit Pa) is expressed by Eq. (11-3), according to the Japanese Industrial Standard (JIS). Now the air temperature in the Kelvin and Celsius scales are denoted as T_a and t_a respectively. The vapor pressure of water in the air p is expressed by Eq. (11-4), where M_v (=18.015 g/mol) is the molar mass of the water vapor and R (=8314.46 J K^{-1} mol^{-1}) is the molar mass constant of the air. By using the standard atmosphere P_0 (=101325 Pa), d_v is expressed by Eq. (11-5). As these equations show, it is

えられる．これらの式からも分かるように，湿度の測定には，温度の測定が必須である．

● 11.2　温度の計測

温度センサは，電気抵抗変化を利用するもの，熱電効果を利用するもの，熱ふく射を利用するものに大別される（Fig. 11-1）．

温度変化に伴う電気抵抗値変化を利用する方法　金属原子は一定の結晶構造を有し，絶対零度以外ではおのおのの平衡位置を中心として格子振動している．式(11-6)に示すのは，電気抵抗 R の温度変化をよく記述する半経験的な式であるグリューナイゼンの式である．Θ はデバイ温度と呼ばれ，結晶の振動の最大角振動数 ω_m を用いて式(11-7)で定義される．k_B はボルツマン定数であり，$\hbar$ はプランク定数 h を 2π で割っ

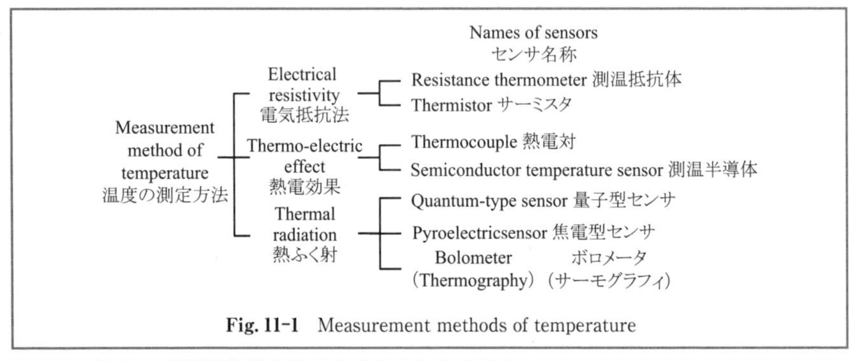

Fig. 11-1　Measurement methods of temperature

necessary to measure temperature T to determine both absolute and relative humidity.

● 11.2　Measurement of temperature

Temperature sensors can be classified into three categories depending on the basis of their measurement principles: electrical resistivity, thermoelectric effect, and thermal radiation.

Thermometry based on electrical resistivity　Metal atoms in lattice vibrate around their equilibrium positions. Eq. (11-6) is referred to as the Bloch-Grueneisen equation and describes the deviation of electrical resistance R associated with deviation in temperature T. The parameter Θ in the equation is referred to as the Debye temperature, which is given by Eq. (11-7). In the equation, ω_m is the

た換算プランク定数である．$\Theta \ll T$ の条件において電気抵抗 R は温度 T に比例するため，電気抵抗値を測定することで温度 T を求めることができる．より簡便には，ほとんどの金属材料の温度 T [K] における抵抗値 R は，温度 T_0 [K] での抵抗値を R_0 として，式(11-8)に示す多項式で与えられ，ある温度範囲においては，二次以上の項を無視して式(11-9)で表せる．式中，α は温度係数である．

■白金測温抵抗体：Fig. 11-2 に示すのは，白金を測温抵抗体として用いた温度センサの例である．セラミック基板上に白金を蒸着して形成した素子にニッケル製のリー

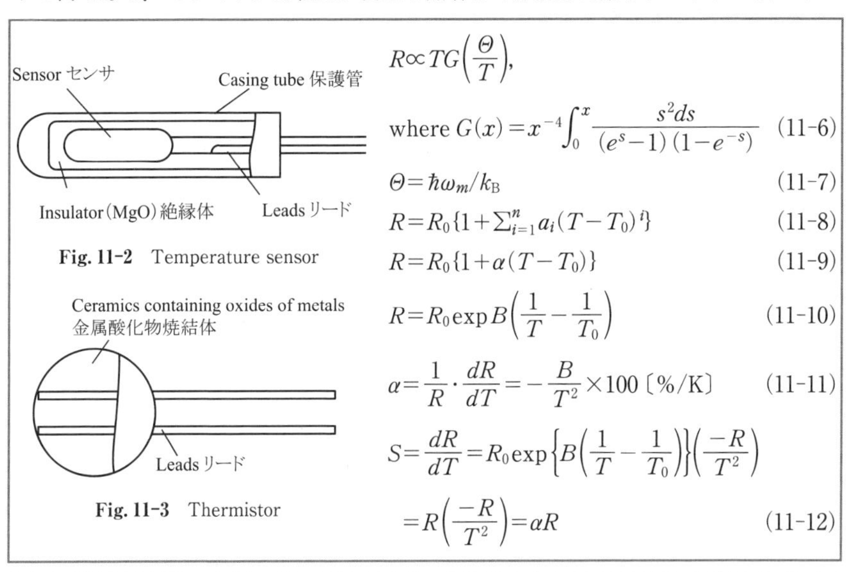

Fig. 11-2 Temperature sensor

Fig. 11-3 Thermistor

$$R \propto TG\left(\frac{\Theta}{T}\right),$$

$$\text{where } G(x) = x^{-4}\int_0^x \frac{s^2 ds}{(e^s-1)(1-e^{-s})} \tag{11-6}$$

$$\Theta = \hbar\omega_m / k_B \tag{11-7}$$

$$R = R_0\{1 + \textstyle\sum_{i=1}^{n} a_i (T - T_0)^i\} \tag{11-8}$$

$$R = R_0\{1 + \alpha(T - T_0)\} \tag{11-9}$$

$$R = R_0 \exp B\left(\frac{1}{T} - \frac{1}{T_0}\right) \tag{11-10}$$

$$\alpha = \frac{1}{R}\cdot\frac{dR}{dT} = -\frac{B}{T^2}\times 100\ [\%/\mathrm{K}] \tag{11-11}$$

$$S = \frac{dR}{dT} = R_0 \exp\left\{B\left(\frac{1}{T} - \frac{1}{T_0}\right)\right\}\left(\frac{-R}{T^2}\right)$$
$$= R\left(\frac{-R}{T^2}\right) = \alpha R \tag{11-12}$$

maximum angular frequency of the lattice vibration, k_B is the Boltzmann constant, and $\hbar$ is the reduced Planck constant, which is the Planck constant divided by 2π. Under the condition of $\Theta \ll T$, T can therefore be detected by measuring electrical resistance R, since R increases in proportion to T. The R of most metallic materials at temperature T [K] can be expressed as in Eq. (11-8), with electrical resistance R_0 at temperature T_0 [K]. Within a limited temperature range, T [K] can be expressed as in Eq. (11-9) by neglecting those terms with orders higher than two. In the equation, α is the temperature coefficient.

■ Platinum resistance thermometer: Figure 11-2 shows an example of a temperature sensor that uses a platinum resistor. A pattern of the sensor made of platinum sputtered on a ceramic substrate is connected to a pair of nickel wires. The electrical resistance of the sensor will be detected by applying a measuring current (2 mA to 10 mA) to the sensor. A highly sensitive temperature measurement can be carried out

ドを取り付けて2～10 mAの電流を流し，電気抵抗値を得る．素子部抵抗値が大きいため導線抵抗の影響が少なく，高感度測定が可能である．

■サーミスタ：Fig. 11-3に，サーミスタの一例を示す．金属酸化物焼結体に白金のリード線を接続している．素子は温度変化に対して大きな電気抵抗値変化を示す．金属とは異なり，式(11-10)で示されるような温度−電気抵抗特性を有している．式中，B [K] は材料および温度に依存するサーミスタ定数であり，一般的には2500～5000 Kの値を有する．温度係数αとBの関係式(11-11)から分かるように，Bが正の場合，αは負となる．また，サーミスタの感度Sは式(11-10)を温度T [K] について微分して式(11-12)のように得られる．

熱電効果を利用する方法　2種類の金属を接合してループを形成すると，両接点の温度差に応じた熱電流が流れることが知られており，この現象をゼーベック効果と呼ぶ．このとき得られる熱起電力は，金属の寸法や形状とは無関係で，金属の種類によってのみ定まる．

with this method, since the influence of the electrical leads are small due to the large electrical resistance of the sensor.

■ Thermistors: Figure 11-3 shows a schematic of a thermistor that consists of a sensor made of metal oxide and a pair of electrical leads made of platinum. Its electrical resistances have a large temperature dependence. Unlike metallic materials, thermistors have a resistance-temperature characteristic described by Eq. (11-10). In the equation, B [K] is referred to as the thermistor coefficient and ranges from 2500 K to 5000 K; the exact value depends on the material and temperature of the thermistor to be used. As Eq. (11-11) shows, in the relationship between the temperature coefficient α and B, α becomes negative when B is positive. By differentiating Eq. (11-10) with respect to T [K], the sensitivity S of the thermistor can be calculated, as shown in Eq. (11-12).

Sensors based on thermoelectric effect　When a closed-loop circuit is established by using two different electrical conductors, electrical current will be generated due to the temperature difference between the two junctions. This phenomenon is referred to as the Seebeck effect. The voltage potential to be generated by the effect is determined only by the materials used in the electrical conductors, regardless of their sizes and shapes.

■熱電対：導線で構成した閉じた輪の一部を熱すると，高温部の電子の運動エネルギー増加に伴い低温部に電子が拡散し，クーロン場が形成される．電子の移動はクーロン力と釣り合う状態で止まり，導線に沿って電位分布が現れる．この釣り合い状態では電流は流れないが，Fig. 11-4 に示すように異なる2つの金属で輪を形成し，片方の接点温度を t_1［℃］，もう片方の接点温度を t_2［℃］とすると，おのおのの金属において温度勾配と電子移動の状態が異なるため，閉回路に電流が流れる．このとき，式(11-13)で示される起電力 E が現れる．$t_2=t_0$，$t_1=0$ とした場合，式(11-14)のとおりとなる．感度 S は式(11-14)を温度 t_0 について微分して式(11-15)のように得られる．熱電対の温度—起電力特性は比較的直線に近く，式(11-15)中の A で近似できることが多い．A はゼーベック係数と呼ばれている．また，この熱電対を多数直列接

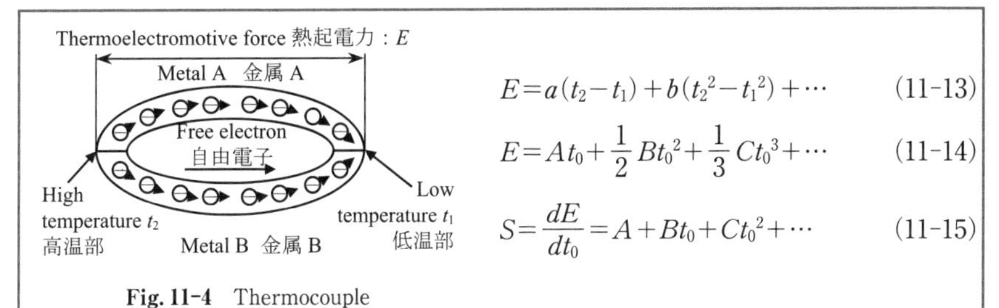

$$E=a(t_2-t_1)+b(t_2^2-t_1^2)+\cdots \quad (11\text{-}13)$$

$$E=At_0+\frac{1}{2}Bt_0^2+\frac{1}{3}Ct_0^3+\cdots \quad (11\text{-}14)$$

$$S=\frac{dE}{dt_0}=A+Bt_0+Ct_0^2+\cdots \quad (11\text{-}15)$$

Fig. 11-4　Thermocouple

■ Thermocouples: When a part of the closed loop is heated, due to the temperature gradient generated, free electrons diffuse to the area with a lower temperature. As a result, a coulomb field will be generated. The diffusion of the electrons will stop as the system reaches an equilibrium state, and a distribution of the voltage potential will also be generated. When the system is in an equilibrium state, electrical current will not flow through the loop. Now, we consider a closed loop consisting of two different conductive metals, as shown in Fig. 11-4. When the temperatures of two junctions are set to be t_1 [℃] and t_2 [℃] respectively, the temperature gradients and the diffusions of the electrons in the two metals will be different. As a result, electrical current will flow through the loop, and an electromotive force E that is calculated by Eq. (11-13) will be generated. When $t_2=t_0$ and $t_1=0$, the equation becomes Eq. (11-14). The sensitivity S can be calculated by differentiating Eq. (11-14) with respect to t_0, as shown in Eq. (11-15). The resistance-temperature characteristics of the thermocouples are almost linear, and can be approximated by the coefficient A, which is referred to as the Seebeck coefficient, as in Eq. (11-15). By connecting several thermocouples in serial, which is referred to as a thermopile, the output voltage can be increased.

続し，出力電圧を高くしたものをサーモパイルという．

ふく射を利用する方法 物体はすべてその温度に対応した波長の電磁波を放出することが知られており，温度 T [K] の物体の放射発散度 W は，シュテファン-ボルツマンの式(11-16)で与えられる．σ はシュテファン-ボルツマン定数である．物体の放射する電磁波（主に赤外線）を検出して温度を測定するのが熱放射式温度センサであり，その方式によって主に量子型，熱型に大別される．量子型センサは吸収したフォトンでキャリアを励起して直接赤外線を検出するもので，動作の原理は第12章に記載の光センサと同一であり，ここでは説明を省略する．熱型センサとしては，焦電センサ，

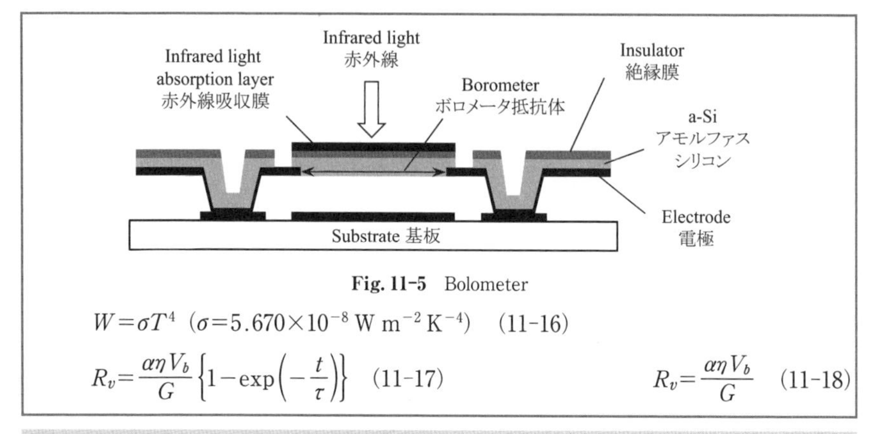

Fig. 11-5 Bolometer

$$W=\sigma T^4 \quad (\sigma=5.670\times10^{-8}\ \mathrm{W\ m^{-2}\ K^{-4}}) \qquad (11\text{-}16)$$

$$R_v=\frac{\alpha\eta V_b}{G}\left\{1-\exp\left(-\frac{t}{\tau}\right)\right\} \qquad (11\text{-}17)$$

$$R_v=\frac{\alpha\eta V_b}{G} \qquad (11\text{-}18)$$

Sensors based on thermal radiation It is widely known that an object emits electromagnetic waves, which have wavelengths corresponding to its temperature, from its surface. The radiant emittance W emitted from an object having temperature T [K] is given by Eq. (11-16), which is referred to as the Stefan–Boltzmann law. In the equation, σ is the Stefan-Boltzmann constant. Sensors based on thermal radiation will measure the temperature of an object by detecting the electromagnetic waves (mainly in the form of infrared light) emitted from the object. These sensors can be categorized into two main types: those based on a thermal detector measuring deviation of its temperature due to radiative energy and those based on an electric current due to the interaction of a photon with an electron. Since the working principle of the latter sensor, which detects infrared light by detecting carriers excited by the photon, is almost the same as that for the light detector in Chapter 12, a detailed description is omitted here. Examples of the former type of sensor include bolometers, pyroelectric sensors and thermopiles.

サーモパイル，ボロメータなどが挙げられる．

■ボロメータ：ボロメータは，赤外線を吸収する吸収材と，それに熱的結合材でつながれた熱浴により構成されており，物体から放射された赤外線を吸収材が吸収した際の温度上昇による抵抗変化を検出する（Fig. 11-5）．電圧感度 R_v は，出力電圧を受光面に入射する赤外線の光量で除した値で定義され，式(11-17)で表される．α はボロメータ抵抗の温度変化率，η は赤外線吸収率，V_b は印加電圧，G は熱コンダクタンス，t は読み出し時間，τ は熱時定数である．読み出し時間が熱時定数に比べて十分大きい場合，R_v は式(11-18)のとおりとなる．ボロメータの原理を応用したセンサがサーモグラフィである．マイクロサイズのボロメータをシリコン基板上に二次元アレイ状に配置したもので，ゲルマニウムのレンズを用いて赤外線を素子上に結像し，温度分布を測定する．

● 11.3 湿度の計測

湿度を計測する方法は，水の気化熱を利用する方法と，感湿素子の電気特性変化を利用するものに大別される．

■ Bolometer: A bolometer is composed of a heat bath and an infrared light absorber connected to the bath. The sensor detects a change in electrical resistance due to the thermoelectric effect induced by the absorbance of infrared light emitted from a measurement target. The voltage sensitivity R_v can be acquired by dividing the output voltage by the intensity of the infrared light created incidental to the active area, as shown in Eq. (11-17). In that equation, α is the temperature coefficient of resistance, η is infrared absorptance, V_b is the voltage supplied, G is thermal conductance, t is readout time, and τ is a thermal time constant. When the readout time is large compared with the thermal time constant, R_v is expressed by Eq. (11-18). One application of a bolometer is as a thermographic instrument. Bolometers fabricated at micrometric size are arranged in a matrix so that temperature distribution of the measurement target can be measured by imaging the infrared light on the bolometers with an image lens made of germanium.

● 11.3 Measurement of humidity

Methods for measuring humidity can be classified into two main categories: those that detect heat in the evaporation of water and those that detect changes in the electrical characteristics of a sensor.

気化熱を利用する方法　空気の湿度が低い場合には水の蒸発量が多く，逆に湿度が高いほど蒸発量は少なくなる．いま，温度計の感温部分を水に濡らしたガーゼなどで覆っておくと（湿球），水分が蒸発するときの気化熱により，雰囲気温度（乾球）よりも ΔT だけ低い温度値を示す．このときの温度低下量は湿度を反映したものであるため，この ΔT をもとに，乾湿計公式から水蒸気圧を求められる．乾球および湿球の2つの温度計から構成されるものを乾湿計（Fig. 11-6）という．乾湿計には，通風しながら測定する通風乾湿計と，通風しない簡易乾湿計とがある．通風乾湿計においては，スプルングの式(11-19)により空気中の水蒸気圧を求める．p は求める水蒸気圧，p_s は湿球温度 T_w における飽和水蒸気圧，A は定数，P は気圧，T は乾球温度である．

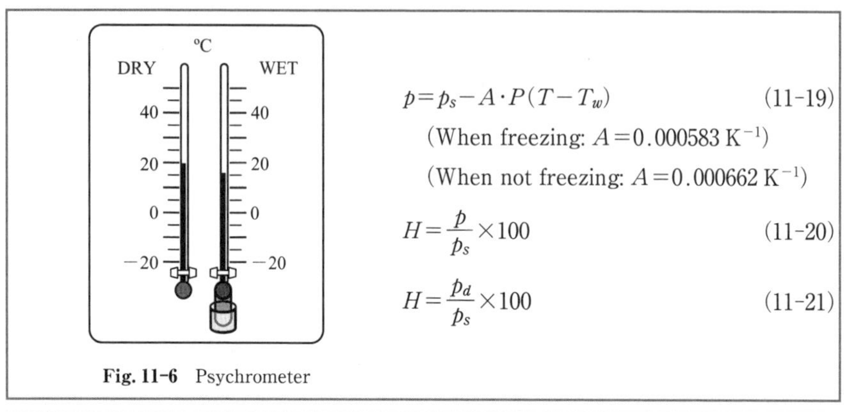

$$p = p_s - A \cdot P(T - T_w) \tag{11-19}$$

(When freezing: $A = 0.000583\ \mathrm{K}^{-1}$)

(When not freezing: $A = 0.000662\ \mathrm{K}^{-1}$)

$$H = \frac{p}{p_s} \times 100 \tag{11-20}$$

$$H = \frac{p_d}{p_s} \times 100 \tag{11-21}$$

Fig. 11-6　Psychrometer

A method that detects heat in the evaporation of water　At a certain temperature, the amount of evaporation will be large under low-humidity conditions, while it will be lower at high humidity. When a temperature sensor is covered with a gauze soaked with water, due to the heat involved in the evaporation of water, the sensor will indicate a temperature with a decrease of ΔT with respect to the atmospheric temperature. Since ΔT reflects ambient humidity, the amount of the water vapor in the air can be obtained from a psychrometer chart. A humidity sensor consisting of wet and dry sensors is referred to as a psychrometer (see Fig. 11-6). There are two kinds of psychrometers: one is aspirated and the other is a simple psychrometer without aspiration. In the case of an aspirated psychrometer, the water vapor pressure in the air is obtained from Eq. (11-19), which is referred to as the Sprung psychrometric formula. In that equation, p is the water vapor pressure to be measured, p_s is the saturated water vapor pressure at the temperature T_w indicated by the wet sensor, A is a constant, P is the ambient pressure, and T is the temperature indicated by the dry sensor. The relative humidity H %RH is obtained

また，相対湿度 H %RH は，式(11-20)から求める．気温が 0℃以下の場合は，湿球が氷結していても，過冷却の水の飽和水蒸気圧を用いる．

感湿素子の特性変化を利用する方法　感湿素子の吸湿および脱湿による特性変化を利用した電子式湿度計には，電気抵抗式，電気容量式などがある．電気抵抗式においては，空気中の水蒸気圧が水の飽和水蒸気圧に等しい温度である露点を測定して水蒸気圧を求める．同時に気温も測定すれば，空気の温度における水の飽和水蒸気圧 p_s，露点における水の飽和水蒸気圧 p_d を用いて式(11-21)より相対湿度を算出可能である．Fig. 11-7(a)に示すのは，塩化リチウム水溶液を用いた露点計である．塩化リチウムは吸湿と脱湿の反応が早く，その飽和水溶液と平衡にある空気の相対湿度が約 11%

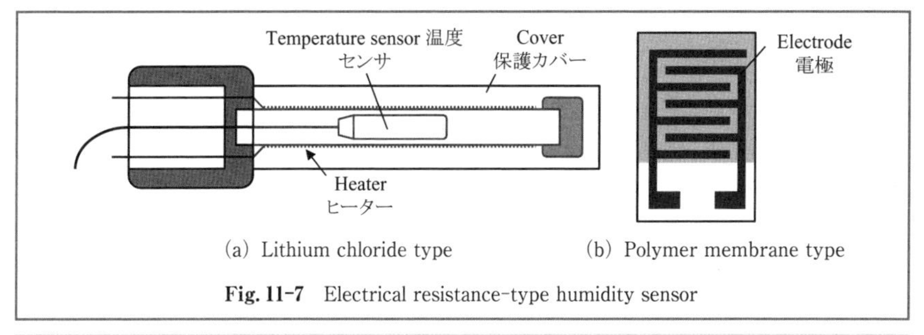

Fig. 11-7　Electrical resistance-type humidity sensor

from Eq. (11-20). In the case of temperatures below 0°C, the saturated water vapor pressure of supercooled water is employed.

A method that detects changes in the electrical characteristics of a sensor　The electrical characteristics of a sensor such as electrical resistance or capacitance, can be used to measure humidity. For a sensor detecting change in electrical resistance, water vapor pressure will be obtained by measuring the dew point, the temperature at which the water vapor pressure p_d in the air is equal to the saturated water vapor pressure p_s. By measuring the ambient temperature at the same time, relative humidity H is obtained from Eq. (11-21) through the use of the detected water vapor pressure p at the ambient temperature. Fig. 11-7(a) shows a schematic of the dew point meter using a solution of lithium chloride in water. Since the lithium chloride absorbs and discharges moisture quickly and the relative humidity of the air in the equilibrium state with the saturated solution is 11%RH, the solution can be employed to measure humidity. When the solution is heated by applying a bias voltage to the electrical lead, the electrical resistance increases at a

RHの低い値を示す特性を利用する．電極線に電圧を印加して水溶液を加熱すると，ある温度で塩化リチウム結晶が析出して電気抵抗が増大する．このときの塩化リチウム膜の水蒸気圧は周囲の空気の水蒸気圧に等しく，この温度を測温抵抗体で検出し，水および塩化リチウム飽和水溶液の温度−水蒸気圧特性表から露点を割り出す．また，導電性高分子膜を用いた露点計では，Fig. 11-7(b)に示すような，くし型などの電極と導電性の高分子膜を表面に形成し，相対湿度が高くなるにつれて増大する感湿部表面を流れる電流を検知する．

【演習問題】

11-1) 27℃は何Kであるか．

11-2) 焦電効果について調べよ．また，焦電型センサの構造と等価回路を示せ．

11-3) 飽和水蒸気圧 $p_s=31.69$ hPa，大気中の水蒸気圧 $p=20.00$ hPa のときの，相対湿度および絶対湿度を求めよ．なお，このときの温度は25℃とする．

11-4) 室温25℃での白金測温抵抗体（温度係数 $\alpha=3.85\times10^{-3}\,\mathrm{K}^{-1}$）の抵抗値を求めよ．なお，0℃での抵抗値を100 Ω とする．

certain temperature with the precipitation of the lithium chloride. At a given moment, the water vapor pressure of the lithium chloride membrane is equal to that of the ambient air. By measuring the temperature in this state, the dew point is obtained by referring to the temperature-water vapor pressure chart of the saturated lithium chloride solution in water.

A dew meter employing a dielectric polymer membrane, a schematic of which is shown in Fig. 11-7 (b), detects an electrical current, which will increase with an increase in the relative humidity, as it flows through the sensor surface.

【Problems】

11-1) What is the value in K of 27℃ ?

11-2) Investigate the pyroelectric effect. In addition, show the schematics of a pyroelectric sensor and its equivalent circuit.

11-3) Obtain the relative humidity and absolute humidity at 25℃ with the saturated water vapor pressure $p_s=31.69$ hPa and the water vapor pressure $p=20.00$ hPa.

11-4) Calculate the electrical resistance of a platinum resistance temperature device (temperature coefficient of resistance $\alpha=3.85\times10^{-3}\,\mathrm{K}^{-1}$) at a room temperature of 25℃. The resistance is 100 Ω at 0℃.

第 12 章　光の計測

光は空間の電場と磁場の変化により形成される電磁波の一種であり，波としての性質と，粒子としての性質をあわせ持ち，工業的にはその波長に応じて異なる呼称で分類されている．本章では，光の強度の計測，および分光計測について述べる．

● 12.1　光の分類

光の波長はおおむね 10 nm から 100 μm の範囲にあり（Fig. 12-1），赤外線より長い波長（100 μm 以上）を有する電磁波は電波，紫外線より短い波長（10 nm 以下）を有する電磁波は X 線（0.01 nm～10 nm）およびガンマ線（0.01 nm 以下）と呼ば

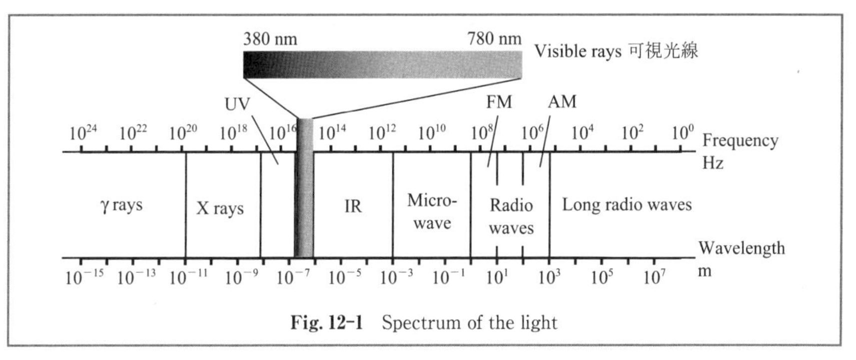

Fig. 12-1　Spectrum of the light

Chapter 12　Measurement of Light

Light is electromagnetic radiation in the electromagnetic spectrum, with properties of both waves and particles. From the industrial point of view, light is categorized by its wavelength. This chapter focuses on light intensity measurement and spectroscopy.

● 12.1　Classification of light

The wavelength of light ranges from 10 nm to 100 μm (see Fig. 12-1). Other electromagnetic waves are categorized as radio waves (longer than 100 μm), X-rays (0.01 nm to 10 nm), gamma rays (shorter than 0.01 nm), etc. A light ray with a

れる．我々人間の目が捕捉可能である波長域 380 nm～780 nm の光は可視光線と呼ばれ，これより短い波長の光を紫外線，長い波長の光を赤外線と呼んで取り扱っている．

光は図 12-2 に示すように，波の進行方向に垂直な横波として正弦波振動して伝搬する電場および磁場として，式(12-1)および式(12-2)で表せる．k は波数で，光の波長 λ を用いて $k=2\pi/\lambda$ と表せる．ω は角速度である．真空中での光の速度を c_0 とすると，屈折率 n の媒質内における速度 c は式(12-3)で表せる．c は λ および周波数 ν を用い，式(12-4)でも表せる．また，光を粒子として扱う場合，光子が有するエネル

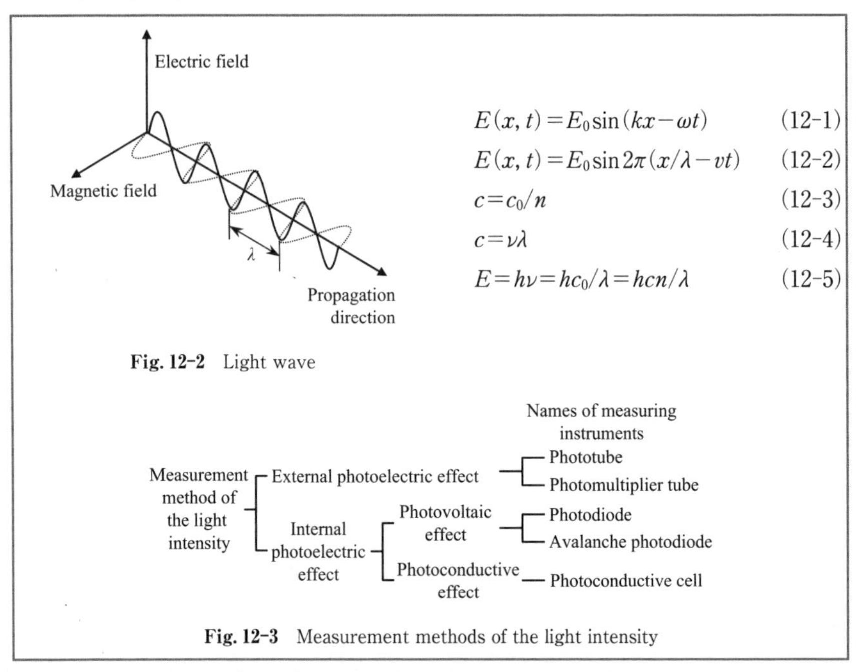

$$E(x,t)=E_0\sin(kx-\omega t) \qquad (12\text{-}1)$$

$$E(x,t)=E_0\sin 2\pi(x/\lambda-vt) \qquad (12\text{-}2)$$

$$c=c_0/n \qquad (12\text{-}3)$$

$$c=\nu\lambda \qquad (12\text{-}4)$$

$$E=h\nu=hc_0/\lambda=hcn/\lambda \qquad (12\text{-}5)$$

Fig. 12-2　Light wave

Fig. 12-3　Measurement methods of the light intensity

wavelength between 380 nm and 780 nm is referred to as visible light. Light rays with shorter wavelengths and with longer wavelengths than visible light are referred to as ultraviolet (UV) and infrared (IR) rays, respectively.

As Fig. 12-2 shows, light can be described as self-propagating transverse oscillating waves of electric and magnetic fields, as expressed in Eqs. (12-1) and (12-2). In those equations, $k=2\pi/\lambda$ is a wave number and ω is angular velocity. The speed of light c in a medium having a refractive index n is expressed in Eq. (12-3), where c_0 is the speed of light in vacuum. The speed of light c is also expressed in Eq. (12-4) by using the light wavelength λ and frequency ν. When light is treated as composed of particles, the energy of light E is calculated as in Eq. (12-5), where $h=6.626\times10^{-27}$

ギー E は式(12-5)で表せる．h はプランクの定数で，$h=6.626\times10^{-27}$ erg である．

● 12.2　光強度の計測

光は電磁波の一種で，波と粒子の性質を有する．この性質が様々な測定に利用されるが，その際に基本となるのが光強度の計測である．この計測に用いられるセンサは，外部光電効果を用いるものと，内部光電効果を用いるものに大別される（Fig. 12-3）．

外部光電効果　金属や固体表面に光が入射した際に，光子が物質と相互作用して電子（光電子）が表面から外部にたたき出される現象が外部光電効果である．いま，振動数 ν の光子のエネルギー E（$=h\nu$）が，固体中のフェルミ準位にある電子を真空中に放出するのに必要なエネルギーの最小値 ϕ_M（仕事関数）より大きい場合，光電子が放出

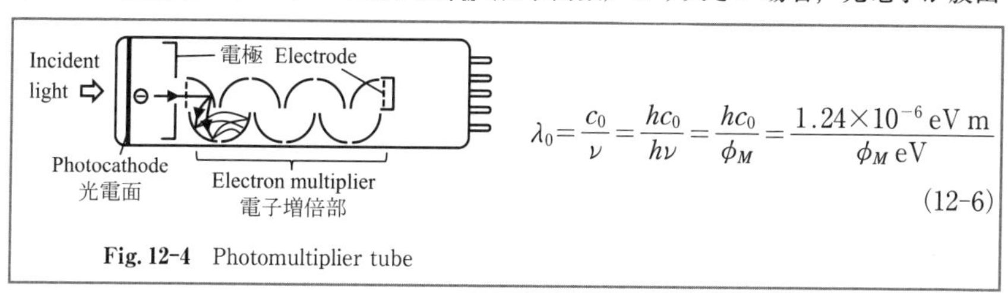

$$\lambda_0=\frac{c_0}{\nu}=\frac{hc_0}{h\nu}=\frac{hc_0}{\phi_M}=\frac{1.24\times10^{-6}\,\mathrm{eV\,m}}{\phi_M\,\mathrm{eV}} \tag{12-6}$$

Fig. 12-4　Photomultiplier tube

erg, the Planck constant.

● 12.2　Measurement of light intensity

The properties of light as waves and particles are utilized in various types of measuring instruments. Measurement of light intensity is fundamental to these measuring instruments. The sensors that measure light intensity can be categorized into two major groups: those based on an external photoelectric effect and those based on an internal photoelectric effect (see Fig. 12-3).

External photoelectric effect　When light rays come into contact with an object surface made of metals or solid materials, electrons (photoelectrons) can be emitted from the surface due to the interaction between the photons and the material. This phenomenon is referred to as the external photoelectric effect. A photoelectron will be emitted when its energy E ($=h\nu$) with a frequency ν is larger than the work function ϕ_M, which corresponds to the minimum energy required to make an electron in the Fermi level emit from the material. The wavelength λ_0 that satisfies $h\nu=\phi_M$ is called the threshold wavelength, and is given by Eq. (12-6).

される．$h\nu=\phi_M$ の場合の光の波長 λ_0 は限界波長と呼ばれ，式(12-6)で与えられる．

■光電子増倍管：光電子増倍管（Fig. 12-4）は，光電管に増倍機能を持たせたものである．光センサの中でもとくに際立った高感度を有し，高速時間応答性に優れる．光が増倍管の端面に配置した光電面に入射すると，光電子が放出される．光電子が加速されて電子増倍部に衝突すると二次電子が放出される．入射光電子に対する二次電子の数量比は電子増倍部の材質および加速電圧に依存するが，多段設置した電子増倍部で電子増幅を繰り返しながら陽極に導き光電流として取り出すことで，光子1個の検出も可能である．

内部光電効果（光起電力効果）　光照射により起電力が発生する現象を光起電力効果という．Fig. 12-5に示すように，P型半導体とN型半導体を接合すると，P型領域

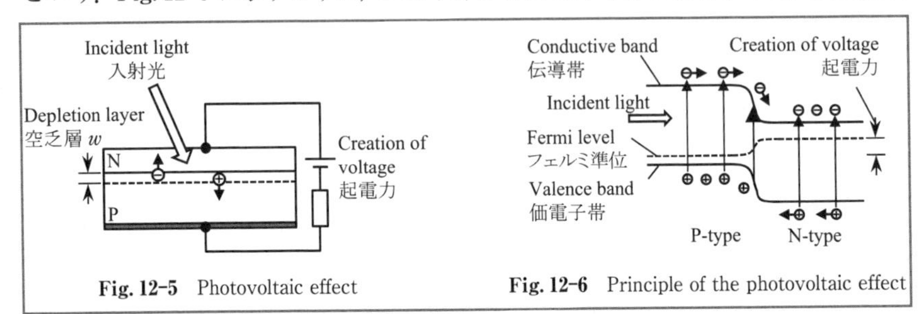

Fig. 12-5　Photovoltaic effect

Fig. 12-6　Principle of the photovoltaic effect

■ Photomultiplier tube: A photomultiplier tube (see Fig. 12-4) has an electron multiplier. It has an extremely high sensitivity among light-detecting sensors and offers a good response in terms of time. When light rays pass through an input window and come into contact with a photocathode, photoelectrons will be emitted from the photocathode. The photoelectrons will be accelerated and focused by the focusing electrode and are multiplied by means of secondary electron emission. A ratio of the number of secondary electrons multiplied by the electron multiplier with respect to the photoelectrons will depend on the material and the acceleration voltage of the electron multiplier. By collecting the multiplied secondary electrons emitted from the electron multiplier, even a single photon can be detected.

Internal photoelectric effect (Photovoltaic effect)　The internal photoelectric effect, which is also referred to as the photovoltaic effect, is the creation of voltage in a material upon its exposure to light. Now, we consider a junction between a P-type semiconductor and an N-type semiconductor, as shown in Fig. 12-5. The

は正孔の濃度が高く，N 型領域は電子の濃度が高いため，正孔は濃度の低い N 型領域へ，電子は濃度の低い P 型領域へ拡散し，接合部に電気的双曲子層（空乏層）が形成される．この層は，電子から見たポテンシャルエネルギーが P 型領域のほうで高くなる傾斜障壁 E_g として作用するため，拡散が抑制される．この P-N 接合部分に光が照射されると，光のエネルギー E が E_g を超える場合には，P 型領域において電子および正孔の対が発生し，伝導帯に励起される．これが内部光電効果である (Fig. 12-6)．励起された電子はポテンシャルエネルギーの傾斜のため N 型領域に移るとともに，正孔は P 型領域に移り，光起電力が発生する．光の照射なしの状態で，P 型領域に N 型領域に対する電位 V を与えると，そのとき発生する電流 I は式(12-

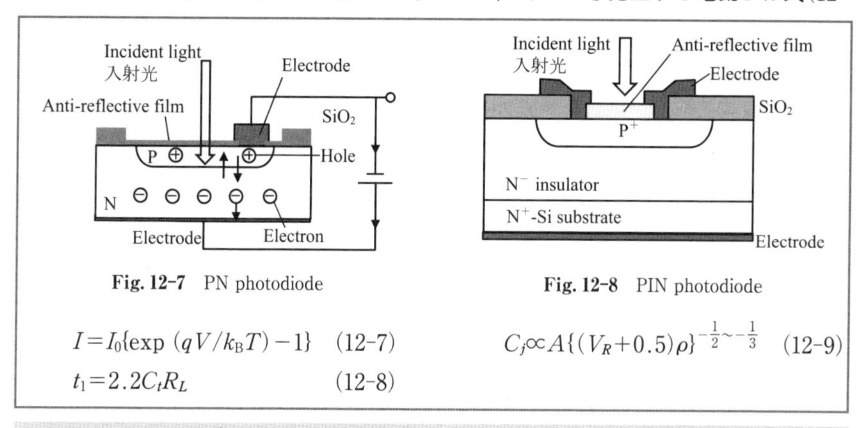

Fig. 12-7 PN photodiode

Fig. 12-8 PIN photodiode

$$I = I_0\{\exp(qV/k_B T) - 1\} \quad (12\text{-}7)$$

$$t_1 = 2.2 C_t R_L \quad (12\text{-}8)$$

$$C_j \propto A\{(V_R + 0.5)\rho\}^{-\frac{1}{2}\sim-\frac{1}{3}} \quad (12\text{-}9)$$

densities of holes (in P-type semiconductors) and electrons (in N-type semiconductors) are high. The holes therefore diffuse from the P-type to the N-type, while the electrons diffuse from the N-type to the P-type. As a result, a layer of the electric dipole, known as the depletion layer, will be generated at the junction. Since this layer acts as an energy barrier E_g, which is higher in the P-type region from the viewpoint of electrons, the diffusion of electrons will be suppressed. When a light ray with energy E higher than E_g is exposed to the P-N junction, a pair of one electron and one hole is generated and is excited into the conductive band in the P-type region. This phenomenon is called the internal photoelectric effect (Fig. 12-6). Due to potential energy, the excited electrons move from the P-type to the N-type, while the poles move from the N-type to the P-type, resulting in the generation of photovoltaic power. The electrical current I generated when an electrical potential V is applied to the P-type region with respect to the N-type region is given by Eq. (12-7). As the equation shows, an electrical current will flow through the junction when the P-type

7)で表される．この式のとおり，P-N 接合はP型領域が正にバイアスされると電流が流れるが，負にバイアスされた場合には，ほとんど電流が流れない，すなわち整流効果を有することが分かる．なお，光検出は一般に $V \leqq 0$ で行われる．

■ Si-PN 接合フォトダイオード/PIN フォトダイオード：Fig. 12-7 に Si-PN 接合フォトダイオードの模式図を示す．N型 Si 単結晶の表面にP型の不純物（ボロンなど）をドープしてP層を構成してP-N 接合とする．P層からの入射光に伴う光電流を検出する．また，P-N 接合の間に真性層（i層）を入れて PIN 構造にしたものを PIN フォトダイオードと呼ぶ（Fig. 12-8）．i層のキャリア濃度は極めて低いため，P層，N層間に逆バイアス電圧を印加するとそのほとんどがi層に印加され，i層は完全に空乏層となる．その結果，P-N 接合界面の空乏層の厚さを大きくした場合と同様の効果を得ることができ，センサの静電容量が小さくなるため，高感度かつ高速なセンシングが可能となる．フォトダイオードの端子間容量を C_t，負荷抵抗を R_L とした場合，その時定数（出力信号が10%から90%に達する時間）t_1 は式(12-8)で表される．

region is biased positive, while very little electrical current will flow through the junction when applying negative bias to the P-type region. This effect is called the rectifying effect. It should be noted that measurement of light is often carried out under the condition of $V \leqq 0$.

■ Si-PN photodiode/PIN photodiode: Fig. 12-7 shows a schematic of a Si-PN photodiode. A P-N junction is prepared by fabricating a P-layer on an N-type monocrystalline silicon surface by doping P-type dopants like boron. The photodiode detects a photocurrent generated by the light rays that illuminate the P-layer. A photodiode with an insulation layer (i-layer) between the P-N junctions is called a PIN photodiode (Fig. 12-8). Since the carrier density in the i-layer is quite low, the reverse bias voltage applied to the P-N junction will be almost directly applied to the i-layer, and the i-layer becomes a complete depletion layer. As a result, the depletion layer has the same effect as the thicker depletion layer between the P-N junctions in the case of a PN diode. Since the capacitance of the PIN photodiode is small, highly sensitive and high-speed light detection can be achieved. The time constant (the time required for the output signal to arise from 10% to 90%) t_1 is given by Eq. (12-8), where C_t and R_L are the capacitance between the electrodes and the load resistance of the photodiode, respectively. A junction capacitance C_j is contained in C_t in proportion to the active area A and is inversely proportional from the second root to the third

C_tに含まれるフォトダイオード接合容量C_jは，受光面積Aにおおよそ比例し，空乏層幅dの二乗根から三乗根に逆比例する．いま，空乏層幅dは逆バイアス電圧V_Rと基板材料の比抵抗ρとの積に比例し，C_jは式(12-9)で表される．したがって，Aが小さくρが大きなフォトダイオードに，逆バイアス電圧を印加すれば応答速度を上げることができる．

内部光電効果（光導電効果）　半導体の持つ禁制帯幅よりも大きいエネルギーを有する光が入射すると，電子・正孔対が発生し，照射部分の導電率が増加する．これを光導電効果という．

■光導電セル：光導電効果を利用したセンサが光導電セルである．Fig. 12-9にその模式図を示す．可視光領域の光検出のため，硫化カドミウム(CdS)セルがよく用いられている．CdSセルは蛇行形とすることで電極との接触面積を大きくとっている．

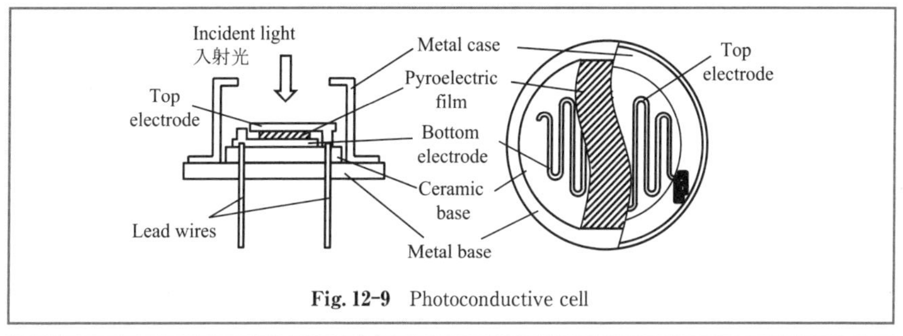

Fig. 12-9　Photoconductive cell

root of the depletion layer width d. Since d is in proportion to the product of the reverse bias voltage V_R and the specific resistance ρ of the substrate, C_j is given by Eq. (12-9). The response speed of the photodiode can be improved by setting a smaller A, a larger ρ, and applying reverse bias voltage.

Internal photoelectric effect (Photoconductive effect)　When light rays with energy greater than the band gap illuminate a semiconductor, electrical conductivity of the illuminated area will be increased with the generation of a pair of an electron and a hole. This phenomenon is called the photoconductive effect

■ Photoconductive cell：A photoconductive cell is a sensor based on the photoconductive effect; Fig. 12-9 shows a schematic of such a sensor. For the detection of visible light, a cell made of cadmium sulfide (CdS) is often employed. The CdS cell, the electrical resistance of which decreases with the increase of intensity of the illuminated light, is often designed in a meandering shape to have a larger contact

CdS セルの抵抗値は照射光強度の上昇とともに低下する．

● 12.3 撮像素子

イメージセンサは，微小な光検出センサをマトリクス状に配置したものであり，CCD 型と CMOS 型がある．

CCD イメージセンサ CCD では，まず，画素内のフォトダイオードにおいて，光電効果を利用して，受光した光を電荷に変換して蓄積する．CCD の画素数だけフォトダイオードを有する．1つの画素から得られる電荷はごく小さく，画像処理に用いるためには増幅する必要がある．そのため，CCD では画素の電荷を隣り合う画素に順次転送して取り出す．Fig. 12-10 にその概略を示す．すべての受光部に蓄積された電荷を，同時に垂直伝送路に転送する．電荷は垂直伝送路を通り，水平伝送路に転送される．水平伝送路を転送されてきた各画素の電荷は，最後の増幅器において電荷から

area with the electrode.

● 12.3 Image sensor

An image sensor consists of small light-detecting sensors aligned in a matrix. Image sensors can be categorized into two main types: charge-coupled devices (CCD) and complementary metal oxide semiconductor (CMOS) devices.

CCD image sensor An electrical charge that corresponds to the amount of illuminated light and is generated by a photoelectric effect at each pixel will be stored in each cell in a CCD. The number of photodiodes in a CCD is the same as the number of pixels. The electrical charge acquired from a single pixel is quite small, so amplification is required to carry out signal processing. Therefore, in CCDs, the electrical charge in a pixel is transferred to the adjacent pixel in sequential order. Fig. 12-10 shows a schematic of the electrical charge transfer. The electrical charge stored in every pixel is transferred to the parallel photodiode shift register simultaneously. The electrical charge goes through the parallel photodiode shift register and is transferred to a serial shift register. The electrode from each pixel is processed by a trans-impedance amplifier placed at the end of the serial shift register. Since all the information in every pixel will be accumulated simultaneously in a CCD device, there

電圧に順次変換・増幅され，信号処理される．各画素の情報が一括で取得されるため，画素間の情報に時間遅延がないのが特徴である．その一方で，水平伝達路，垂直伝達路を介して各画素の情報を順次読み取るため，信号処理に時間を要する．

CMOS(相補形金属酸化膜半導体)**イメージセンサ**　CMOS イメージセンサは，PN フォトダイオードに，金属と半導体の間に薄い酸化膜が挟まれた MOS 構造 (Fig. 12-11) により形成したトランジスタを組み合わせた撮像素子である．一般的な集積回路半導

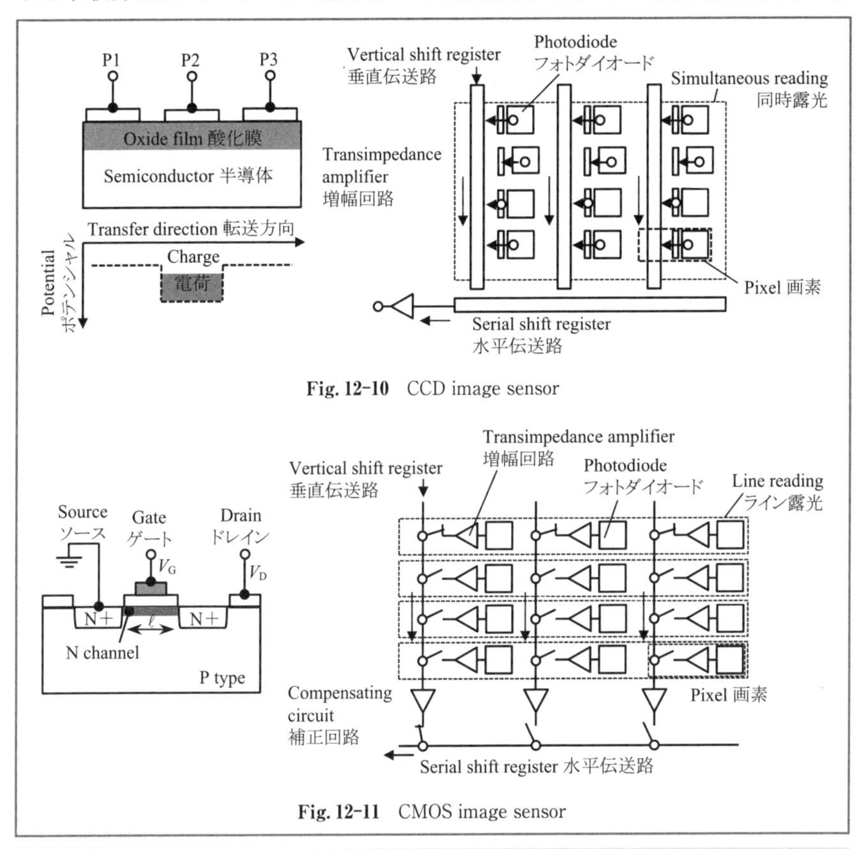

Fig. 12-10　CCD image sensor

Fig. 12-11　CMOS image sensor

is no time delay in obtaining the information from every pixel. However, there is a relatively long processing time, since the electrical charge is read in the sequential order described above.

CMOS (Complementary metal oxide semiconductor) **image sensor**　A CMOS image sensor is composed of PN photodiodes and transistors with MOS structures,

体プロセスによる素子形成が可能である．PN フォトダイオードの各画素に CMOS 増幅回路が付与されている．光電効果を利用して蓄積した電荷は画素内の増幅回路により電圧に変換および増幅される．増幅された電圧は，行ごとに垂直伝送路に転送される．その後，垂直伝送路ごとに配置されている列回路により画素間のバラツキを補正した後，水平信号線に送られる．ラインごとに画素情報を読みとることが可能であり，CCD に比べて情報転送速度が速いことが特徴である．

● 12.4　光スペクトルの計測

太陽光や白熱電球からの光をガラスプリズムに通すと，プリズム中の光の屈折率は短波長になるほど大きくなる性質があるため，結果として波長により分散（分光）させることができる．このような性質を利用してある特定波長の光を選択する装置を，分光器，あるいはモノクロメータと呼ぶ．一般的な計測機器の分光には，回折格子がよく用いられる．Fig. 12-12 は，透明ガラス基板上に等間隔のラインを蒸着した透過

which consist of a metal oxide layer sandwiched between a metal layer and a semiconductor layer (see Fig. 12-11). The sensor can be fabricated by using an ordinary semiconductor fabrication process for integrated circuits. Each PN photodiode is equipped with a CMOS amplifier circuit in each pixel. The electrical charge generated by the photoelectric effect and accumulated in each pixel will be processed by a transimpedance amplifier. The amplified voltage output from each pixel will be transferred to the vertical shift register for each column. After that, the electrical circuits arranged in each column will compensate for the deviation of each pixel and the electrical signal will be transferred to a serial shift register. Since the information for each line is acquired simultaneously, the information transfer rate is faster than that of a CCD image sensor.

● 12.4　Measurement of light spectrum

Since a refractive index in a glass increases with a decrease in light wavelength, light emitted from the sun or a filament lamp can be dispersed by a prism. Measuring instruments utilizing this phenomenon to extract light with a specified wavelength are called as spectroscopes or monochromators. For commonly used instruments, diffraction gratings are often employed. Fig. 12-12 shows a schematic of a transmission diffraction grating that has slits with a constant period sputtered on a

型回折格子の例である．いま，波面が平行な光（波長λ）が，間隔dのスリットを有する回折格子に対して角度αで入射した場合を考える．隣り合うスリットを通過する光線間の光路差OPDは，出射角をβとして式(12-10)で表される．式(12-11)に示すように，このOPDが波長の整数倍と等しい場合，2本の光線が強めあう．このときの出射角β_mは式(12-12)のとおり得られる．この式より，距離L（$\gg d$）だけ離れたスクリーンに生じる干渉縞の周期（回折光の間隔）Dは，式(12-13)のとおり得

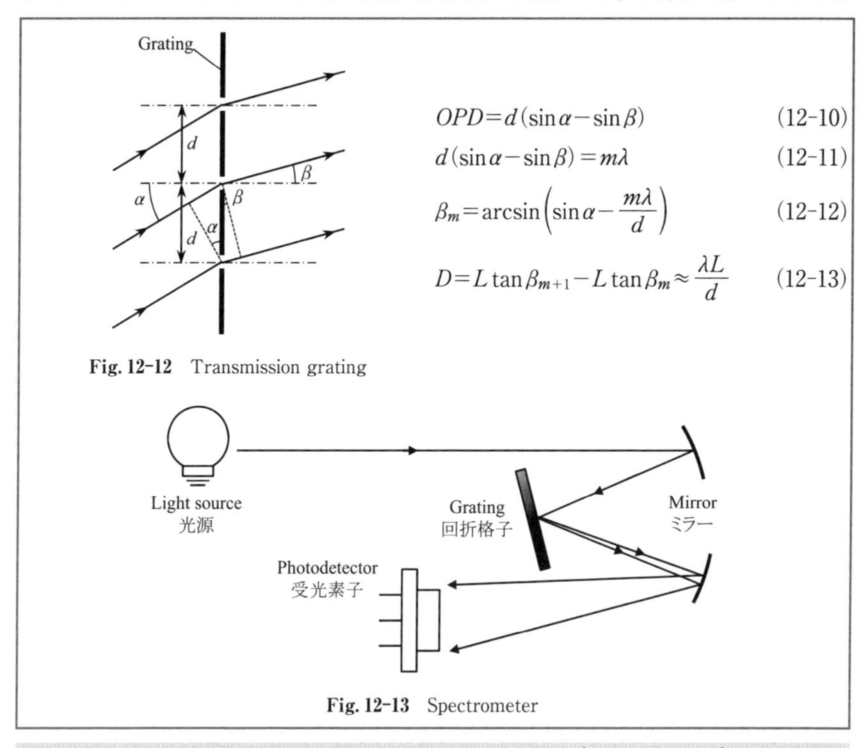

$$OPD = d(\sin\alpha - \sin\beta) \qquad (12\text{-}10)$$

$$d(\sin\alpha - \sin\beta) = m\lambda \qquad (12\text{-}11)$$

$$\beta_m = \arcsin\left(\sin\alpha - \frac{m\lambda}{d}\right) \qquad (12\text{-}12)$$

$$D = L\tan\beta_{m+1} - L\tan\beta_m \approx \frac{\lambda L}{d} \qquad (12\text{-}13)$$

Fig. 12-12　Transmission grating

Fig. 12-13　Spectrometer

glass plate. Now, we consider a light with a plane wave (wavelength λ) that comes into contact with a grating that has a grating period d and an angle of incidence α. The optical path difference (OPD) of adjacent light rays is given by Eq. (12-10) by using the angle β. When the OPD is equal to the integral multiple of the light wavelength, as shown in Eq. (12-11), the light rays will produce constructive interference. The angle β_m at the point of constructive interference is given by Eq. (12-12). From the equation, the period of the interference fringes (the distance between the neighboring diffracted beams) D on a screen at a distance L ($\gg d$) from the grating is acquired by Eq. (12-13). It should be noted that this equation can also be applied to cases with a

られる．なお，これらの式は反射型の回折格子にも適用できる．

回折格子を用いた分光計の一例を Fig. 12-13 に示す．光ファイバからの光をスリットを通して導入し，コリメートミラーで平行光にした後，回折格子で分光する．その後，フォーカスミラーにより回折格子で分光した光をフォトディテクタ上に結像する．光の波長により異なるフォトディテクタ上での結像位置を検出することで，光学系に導入された光のスペクトラムを得ることができる．

【演習問題】

12-1） フォトダイオード接合容量 C_j が受光面積 A に比例し，かつ空乏層幅 d の二乗根に逆比例するものとして，$A=1\ \mathrm{mm}^2$，$\rho=10^3\ \Omega/\mathrm{m}$ の条件のもと，逆バイアス電圧 5 V の引加の有無により，どの程度応答速度が上がるか，計算せよ．

12-2） 波長 $\lambda=675\ \mathrm{nm}$ のコリメートレーザ光を格子間隔 4 μm の回折格子に垂直に入射した場合の，一次回折光の回折角を求めよ．

12-3） CCD および CMOS センサの出力信号におけるノイズレベルについて調べ，比較せよ．

reflective grating.

Fig. 12-13 shows an example of a spectroscope employing a diffraction grating. A light ray from an optical fiber is guided into the optical setup. The light collimated by a collimating mirror comes into contact with the diffraction grating to disperse the light. The dispersed light is focused onto a photodetector by using a focusing mirror. By detecting the position of the focused light on the photodetector, the optical spectrum of the light guided into the spectroscope can be evaluated.

【Problems】

12-1) Assume that the junction capacitance C_j of a photodiode is in proportion to the active area A and is inversely proportional to the second root of a depletion layer of width d. Calculate the increase in response speed of the photodiode with the existence of a reverse bias voltage of 5 V under the condition that $A=1\ \mathrm{mm}^2$ and $\rho=10^3\ \Omega/\mathrm{m}$.

12-2) Obtain the diffraction angle of the first-order diffracted beam when a collimated laser beam with a wavelength λ of 675 nm comes into perpendicular contact with the grating that has a grating period of 4 μm.

12-4)　反射型回折格子におけるリトロー配置について調べよ．

12-5)　光スペクトラムを利用した測定装置を挙げよ．

12-3)　Investigate the noise level of output signals from CCD and CMOS sensors respectively.

12-4)　Investigate the Littrow configuration of reflective-type diffraction grating.

12-5)　Indicate appropriate measuring instruments based on the optical spectrum.

第 13 章　電気磁気の計測

種々のセンサの出力は電気信号である．本章では，電圧，電流，インピーダンスの測定方法について説明し，空間に発生する電界および磁界の検出方法を述べる．

● 13.1　電気と磁気の関係

電気と磁気の計測では，直流と交流の計測が行われる．直流の計測では，時間的に変化しない直流回路の電圧，電流，空間に発生した静電界，静磁界などが計測対象となる．交流の計測では，電圧，電流は周期的に変化するので，電圧振幅，電流振幅に

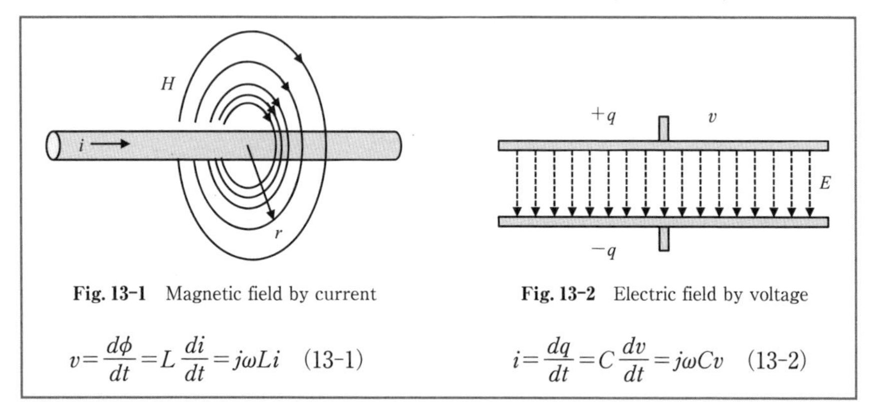

Fig. 13-1　Magnetic field by current

$$v=\frac{d\phi}{dt}=L\frac{di}{dt}=j\omega Li \quad (13\text{-}1)$$

Fig. 13-2　Electric field by voltage

$$i=\frac{dq}{dt}=C\frac{dv}{dt}=j\omega Cv \quad (13\text{-}2)$$

Chapter 13　Measurements of Electricity and Magnetism

Outputs of sensors are often electrical signals. In this chapter, measurement methods for voltage, current, and impedance are explained and detection methods of electric and magnetic fields in space are described.

● 13.1　Relationship between electricity and magnetism

Direct current (DC) and alternating current (AC) are detected in the measurements of electricity and magnetism. In DC measurements, voltage, current, and electrostatic and magnetic fields generated in space, etc., which do not vary temporally, are measured. In AC measurements, as voltage and current vary

加えて，それらの位相差が求められる．電流 i は抵抗 R に電圧 v を発生させる．この関係 $v=Ri$ はオームの法則である．このとき抵抗の長さを l とすると，単位長さあたりに v/l の電界 E が発生する．電圧による電位差が生じている場所には，電界 E が常に発生している．

電流 i がコイルに流れるとアンペールの法則により，Fig. 13-1 のように磁界 $H=i/(2\pi r)$ が発生する．ここで r は半径である．磁束密度 B は $B=\mu H$ で与えられる．ここで μ は媒質の透磁率である．また，回路に鎖交する磁束 ϕ はインダクタンス L を用いて $\phi=Li$ と表現される．交流では磁束の時間変化が生じ，電磁誘導の法則により，式(13-1)で表される電圧 v が発生している．ここで交流の角周波数を ω とした．電圧の大きさは $|v|=\omega L|i|$ となる．電流が流れている領域の周囲には，磁界が発生し，磁束の時間変化により交流電圧が発生する．

電圧 v がコンデンサに印加されると，コンデンサの対向電極には電荷 q が蓄積され，$q=Cv$ となる．ここで C は静電容量である．電荷 q は電流の蓄積であるので，電流は式(13-2)により電荷の微分で与えられる．電流の大きさは $|i|=\omega C|v|$ とな

periodically, the phase difference between them is measured, along with the amplitude. Current i generates voltage v across resistor R. This relation $v=Ri$ is called Ohm's law. Then, the electric field E equal to v divided by the length l of the resistor is generated per unit length. The electric field is always generated in the place where a potential difference in voltage exists.

When current i flows across a coil, a magnetic field $H=i/(2\pi r)$ is generated by Ampere's law, as shown in Fig. 13-1, where r is a radius. Magnetic flux density B is given by $B=\mu H$, where μ is the permeability of a medium. Interlinked magnetic flux ϕ is given by $\phi=Li$ using the inductance L of a circuit. The time variation of ϕ caused by an AC current generates a voltage v by electromagnetic induction, as shown in Eq. (13-1). Here, the angular frequency is ω. The magnitude of voltage is given by $|v|=\omega L|i|$. A magnetic field is generated around the flowing current and AC voltage is generated by the time variation of the magnetic flux.

When a voltage v is applied to a capacitor, charge q is accumulated on the facing electrodes of the capacitor and $q=Cv$, where C is capacitance. As the charge q is the accumulation of current, the current i is given by differentiating the charge, as expressed in Eq. (13-2). The magnitude of the current is given by $|i|=\omega C|v|$. An electric field E is generated between the two electrodes of the capacitor, as shown in

る．2つの電極の間には Fig. 13-2 のように電界 E が発生する．平行平板電極の場合は電極間隙を d とすると，発生する電界は $E=v/d$ により与えられる．

● 13.2　電気量の計測

電流の計測　電圧，電流の計測にはそれぞれ電圧計，電流計を用いるが，精度のよい計測を行うためには，計測対象に与える影響をできるだけ少なくする必要がある．電流計の内部抵抗が計測に与える影響を考えてみよう．メーターのように指針の振れにより測定する方式では，流れる電流の磁界による力で振れを発生する．電流計を用いるとき，計測したい回路の一部を切断し，電流計を接続する．電流計から接続した回路を見ると，等価抵抗と電源で表せるので，Fig. 13-3 に示すように電流計の内部抵抗 r_A と外部回路の等価抵抗 R は電源に直列となる．流れる電流は $i=v/(R+r_A)$ と

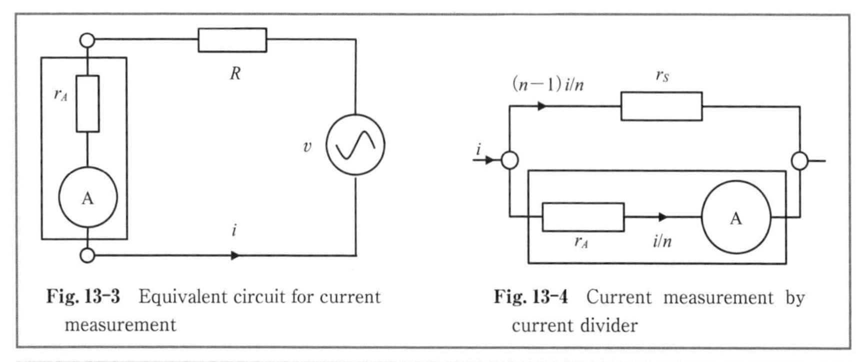

Fig. 13-3　Equivalent circuit for current measurement

Fig. 13-4　Current measurement by current divider

Fig. 13-2. The electric field generated is given by $E=v/d$ in the case of parallel plate electrodes with a gap d.

● 13.2　Electric measurement

Measurement of current　In the measurement of voltage and current, voltmeters and ammeters are used. In order to obtain high measurement precision, it is important to minimize the influence of the measurement on the measuring object. Let us consider the influence of the internal resistance of an ammeter on the measurement. In an ammeter, the deflection of the ammeter needle is generated by the electromagnetic force of a current. The ammeter is inserted between disconnected terminals. A two-terminal circuit is expressed as a series connection of an equivalent resistor R and a voltage source v, as shown in Fig. 13-3. The inner resistance r_A of the ammeter and R are connected in series. The current is given by $i=v/(R+r_A)$, which is smaller than the current ($i=v/R$) before connecting the ammeter. The value of r_A

なり，電流計を入れる前に抵抗 R に流れている電流（$i=v/R$）より小さくなる．R に対して r_A はできるだけ小さいほうがよい．大きい電流を計測する場合には，振り切れるので，分流器により分流する．このとき，分流器に大部分の電流を流し，一部の電流を適切に電流計に流すには，分流器の抵抗は電流計の内部抵抗 r_A より，十分小さくする必要がある．Fig. 13-4 において，分流器（抵抗 r_S）に流れる電流と内部抵抗 r_A に流れる電流の比が $n-1:1$ であるためには，$r_S=r_A/(n-1)$ となり，n が大きい場合，極めて小さい抵抗が必要となる．もし電流計がない場合は，回路に小さい抵抗を挿入し，その抵抗に発生する電圧を電圧計で測定し，電流を求めることもできる．

電圧の計測　電流の流れている回路の一部の素子（ここでは抵抗 R）に発生している電圧 v を測定する．電圧計は大きい抵抗 r_v と電流計で構成できる．電圧計が計測に与える影響について考えよう．最も簡単な例として，Fig. 13-5 のように電流源 i

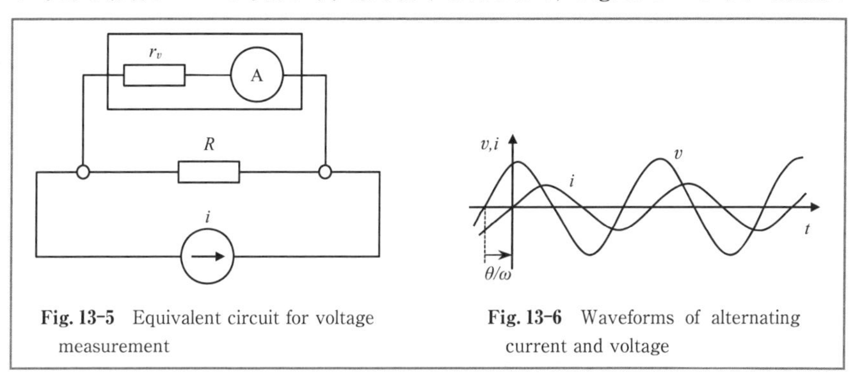

Fig. 13-5　Equivalent circuit for voltage measurement

Fig. 13-6　Waveforms of alternating current and voltage

should be much smaller than R. When measuring a large current, the current is split by using an electrical shunt so as not to exceed the scale. The resistance r_S of the shunt should be much smaller than the internal resistance r_A so that the most of the current flows into the shunt and only a small potion of the current flows into the ammeter. In Fig. 13-4, when the ratio between the currents flowing in r_S and r_A is equal to $n-1:1$, the relation $r_S=r_A/(n-1)$ is obtained. A very small resistance is needed when n is large. If an ammeter is not available, a voltmeter can be used to measure the current across a small resistance that is inserted in the circuit loop.

Measurement of voltage　Here, we consider the measurement of voltage generated across circuit elements (a resistor). A voltmeter consists of a large resistor r_v and an ammeter. Let us consider the influence of the voltmeter on the

に抵抗 R が接続されており，抵抗の両端の電圧を測定する場合を考える．電圧計の指示は r_v に流れる電流を加えて，$v=iR/\{1+(R/r_v)\}$ となるので，電圧計を接続しない場合の電圧 $v=iR$ に比べて小さい値となる．$R \ll r_v$ の場合は，電圧計の接続による変化は無視できる．また，電圧計の測定範囲を拡大するため倍率器を用いる．倍率器は大きい抵抗で，電圧計に直列につなぎ，電圧計にかかる電圧を分圧し，電流計に流れる電流を最大値より小さくする．

インピーダンスの計測 インピーダンスは交流電圧と流れる交流電流の比で定義できる複素数量（ベクトル量）である．インピーダンス $\boldsymbol{Z}$ は抵抗 R とリアクタンス X により $\boldsymbol{Z}=R+jX$ と表現できる．リアクタンスはコンデンサの容量性成分 $X_C=-1/(\omega C)$ とコイルの誘導性成分 $X_L=\omega L$ に分類でき，負の値の場合は容量性となり，正の値の場合は誘導性となる．また，インピーダンス $\boldsymbol{Z}$ は電圧と電流の比であるので，$\boldsymbol{Z}=\boldsymbol{V}/\boldsymbol{I}=|\boldsymbol{V}/\boldsymbol{I}|\exp(j\theta)$ と表され，Fig. 13-6 に示すように交流電圧と電流の波形より，電流に対する電圧の位相差 θ が求められる．θ が正のときは誘導性で，負のときは容量性のインピーダンスとなる．最も簡単には，電流と電圧の正弦波の波形をオシ

measurement. As a simple example, a current source i is connected to a resistor R, as shown in Fig. 13-5, and the voltage across R is measured. The output of a voltmeter is given by $v=iR/\{1+(R/r_v)\}$, adding a current that flows across r_v, which is smaller than without the voltmeter. When $R \ll r_v$, the influence of the voltmeter is negligible. To increase the measurement range, a multiplier is used. The multiplier consists of a large resistor that is connected in series with the voltmeter so that the voltage applied to the voltmeter is decreased.

Measurement of impedance Impedance is the complex (vector) value defined by AC voltage divided by alternating current. The impedance $\boldsymbol{Z}$ is expressed by $\boldsymbol{Z}=R+jX$ with a resistance R and reactance X. The reactance is classified into a capacitive component $X_C=-1/(\omega C)$ and an inductive component $X_L=\omega L$. When the value X is negative, the reactance is capacitive. When X is positive, it is inductive. The impedance $\boldsymbol{Z}$ is expressed by $\boldsymbol{Z}=\boldsymbol{V}/\boldsymbol{I}=|\boldsymbol{V}/\boldsymbol{I}|\exp(j\theta)$. As shown in Fig. 13-6, the phase difference θ is obtained from the waveforms of the voltage and the current. When θ is positive, the reactance is inductive, and when negative, it is capacitive. In the simplest case, voltage and current are observed by an oscilloscope and the

ロスコープで観測し，振幅と位相差よりインピーダンスの値を求めることができる．

不明であるインピーダンスは，しばしば，第14章で説明されるホイートストンブリッジ回路を用いて求められる．3つのインピーダンスの内，少なくとも1つは可変のインピーダンス（空気間隙のコンデンサおよび可変抵抗）であり，中央の検流計の値がゼロになるよう調整し，ブリッジ条件より求める．

インピーダンス計測法のなかで，共振法は，主に誘導性のインピーダンスの測定に用いられる．Fig. 13-7 において，流れる電流 $\boldsymbol{I}$ は $\boldsymbol{I}=\boldsymbol{V}_S/[R_X+j\{\omega L_X-1/(\omega C)\}]$ である．発振器の周波数を変えて，不明のインピーダンスの誘導性リアクタンス成分 ωL が可変コンデンサの容量性リアクタンス成分で打ち消されるとき，共振状態が発生する．すなわち共振角周波数 $\omega_0=\{1/(L_XC)\}^{1/2}$ のとき，電流は最大となり，可変コンデンサの電圧 V_C は極大値をとる．このとき，共振角周波数 ω_0 と静電容量 C の値から，L_X の値を決定できる．

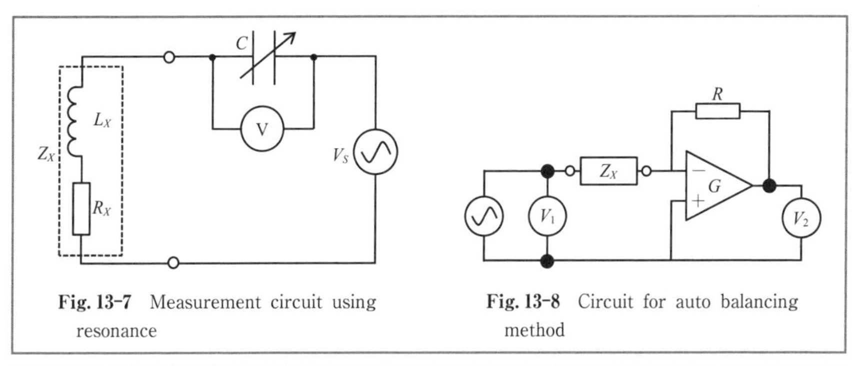

Fig. 13-7　Measurement circuit using resonance

Fig. 13-8　Circuit for auto balancing method

impedance is obtained from the amplitudes and the phase difference.

Unknown impedance is often determined by a Wheatstone bridge circuit, detailed in Chapter 14. If at least one of three impedances of the bridge is variable (air-gap condenser or variable resistor), the balanced condition of the bridge is obtained from the zero current condition of an ammeter placed in the center.

Impedance measurement using the resonance phenomenon is used mainly for measuring inductive impedance. In Fig. 13-7, the flowing current $\boldsymbol{I}$ is given by $\boldsymbol{I}=\boldsymbol{V}_S/[R_X+j\{\omega L_X-1/(\omega C)\}]$. By varying the frequency of the oscillator, the reactance component ωL_X of unknown impedance is compensated for by the capacitive reactance of a variable capacitor. A resonant condition is then generated, in the resonant angular frequency condition $\omega_0=\{1/(L_XC)\}^{1/2}$. The current is maximized and the voltage V_C across the variable capacitor becomes the maximum value. Under this condition, the value of L_X is determined from ω_0 and C.

自動平衡ブリッジ法においては，Fig. 13-8 に示すように，高い増幅率の差動増幅器を用いる．不明なインピーダンス $\boldsymbol{Z}_X$ を増幅器の反転入力端子に接続し，帰還抵抗 R につなげる．入力電圧 $\boldsymbol{V}_1$ をインピーダンス $\boldsymbol{Z}_X$ の他端子に印加する．増幅器の反転入力端子は仮想接地（電位 0 V）になるよう増幅率が自動的に調整されるので，帰還抵抗 R に発生する電圧 $\boldsymbol{V}_2$ と $\boldsymbol{V}_1$ の比より不明インピーダンスは $\boldsymbol{Z}_X=R\times\boldsymbol{V}_1/\boldsymbol{V}_2$ より求められる．このとき電圧 $\boldsymbol{V}_1$，$\boldsymbol{V}_2$ は位相角も含めて知る必要があり，$\boldsymbol{V}_1=|\boldsymbol{V}_1|\exp(j\theta_1)$，$\boldsymbol{V}_2=|\boldsymbol{V}_2|\exp(j\theta_2)$ と表され，$\boldsymbol{Z}_X=R|\boldsymbol{V}_1|/|\boldsymbol{V}_2|\exp\{j(\theta_1-\theta_2)\}$ となる．

センサと計測機が離れている場合，Fig. 13-9 に示すように，伝送線路（ケーブル）により信号を伝送する．とくに高周波の測定では，伝送線路の特性インピーダンスと信号源インピーダンスおよび計測機の入力インピーダンスが整合すると，信号の反射波が発生せず，信号の電力を最もよく伝送できる．この条件は各抵抗成分が等しく，

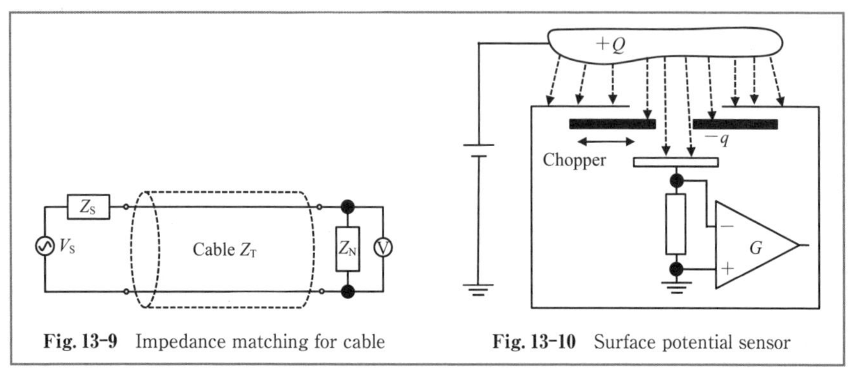

Fig. 13-9 Impedance matching for cable　　**Fig. 13-10** Surface potential sensor

In the auto-balancing method, a high-gain differential amplifier is used, as shown in Fig. 13-8. The unknown impedance $\boldsymbol{Z}_X$ is connected to the inverting input terminal and one end of feedback resistance R. An input voltage $\boldsymbol{V}_1$ is applied to the other terminal of $\boldsymbol{Z}_X$. Since the gain of the amplifier is automatically adjusted so as to "ground" the inverting input terminal (zero volts), the unknown impedance is found from the voltage $\boldsymbol{V}_2$ across the feedback resistance R and $\boldsymbol{V}_1$, using the equation $\boldsymbol{Z}_X=R\times\boldsymbol{V}_1/\boldsymbol{V}_2$. In this measurement, the voltages $\boldsymbol{V}_1$ and $\boldsymbol{V}_2$ should be determined through their phase angles: $\boldsymbol{V}_1=|\boldsymbol{V}_1|\exp(j\theta_1)$, $\boldsymbol{V}_2=|\boldsymbol{V}_2|\exp(j\theta_2)$, and $\boldsymbol{Z}_X=R|\boldsymbol{V}_1|/|\boldsymbol{V}_2|\exp(j(\theta_1-\theta_2))$.

When a sensor and measuring instrument are separated, a signal is often transmitted through a cable, as shown in Fig. 13-9. Especially in measurements at high frequencies, signal reflections at the interfaces between the cable and the sensor and measuring instrument are not generated and the signal power is most efficiently transmitted when the signal impedance and the input impedance of the measuring

各リアクタンス成分がゼロになる場合である.

電界測定 直流の電界のセンサとして，Fig. 13-10 に示す表面電位センサがある．帯電物体の電位を，静電誘導により，検出電極に発生した電荷から計測する．物体が直流電圧で帯電している場合，検出電極には直流電荷が発生する．この電荷は小さく，直流的なノイズの影響を受けやすい．検出電極の前に，電界の侵入を部分的に遮る金属振動板を設置し，振動させることで，誘導された電荷を振動周波数の交流に変換し，感度良く検出できる．

Fig. 13-11 に示す電気光学効果を用いた電界測定では，誘電体の探針を用いる．光学的に測定できるので，計測対象に与える擾乱（じょうらん）は少ない．また，電気光学効果は直流から電波の高い周波数の交流まで応答するので，広い帯域で計測できる．電気光学

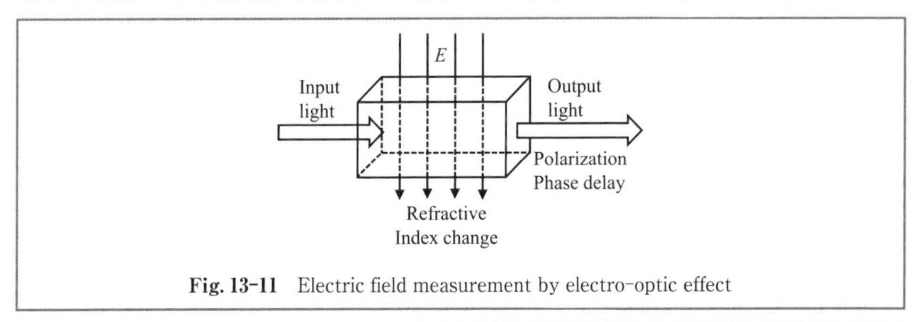

Fig. 13-11 Electric field measurement by electro-optic effect

instrument match the characteristic impedance of the cable. Under these conditions, the resistances of the respective impedances are all equal and the reactances of the impedances are zero.

Electric field measurement The surface potential sensor shown in Fig. 13-10 is known as a DC electric field sensor. The potential of a charged object is measured from the charge on a sensing electrode, which is generated by electrostatic induction. When the object is charged with a DC voltage, a DC charge is generated on the sensing electrode. The charge is often very small and easily affected by DC noise. A metal plate is placed before the sensing electrode and oscillated to block the electric field periodically. Then, the induced charge is converted to alternating current and detected with good sensitivity.

In electric field measurement using the electro-optic effect (EOE), as shown in Fig. 13-11, a dielectric probe is used. Because optical sensing is used, the electrical influence is small. In addition, EOE occurs with DC and high frequencies. EOE is an

効果は，誘電体結晶に外部から電場を加えると，光学的な複屈折を生じる現象である．すなわち，結晶の屈折率が変化するので，その中を通る光に対して，位相変化を発生させる．光の位相変化を偏光変化，干渉強度の変化などより求め，電界強度を導出できる．電気光学効果には屈折率変化が電界強度の一乗に比例する一次電気光学効果（ポッケルス効果）と二乗に比例する二次電気光学効果（カー効果）がある．電界センサには一次の効果が主に用いられる．応答時間は 0.01 ps～0.1 ps である．材料としては，$LiTaO_3$，$LiNbO_3$，CdTe，KTP，BSO などの誘電体結晶材料が用いられる．感度は 1 V/m 程度である．

13.3　磁気量の計測

コイルによる鎖交磁束計測　磁気量の計測では，磁界 H あるいは磁束密度 B を測定する．$B=\mu H$ の関係があるので，どちらかを知ればよい．ここで μ は透磁率である．コイルに鎖交する磁束 ϕ が時間的に変化すると，電磁誘導により，コイルに電圧が発生する．発生電圧 v は $v=-d\phi/dt$ で求められる．磁束 ϕ が交流的に変化しておれば $\phi=\phi_0\exp(j\omega t)$ として，$v=-j\omega\phi_0\exp(j\omega t)$ であるので，電圧と周波数から鎖交

optical birefringence generated by the voltage applied to the dielectric crystal. The refractive index change of crystal causes a phase change in the propagating light. The phase change is detected from either the polarization change or the interference intensity change. EOEs are categorized into two kinds: first order EOEs (employing the Pockels effect), where the refractive index change is proportional to the magnitude of the electric field and second order EOEs (employing the Kerr effect), where the refractive index change is proportional to the square of the magnitude of the electric field. The first order effect is used mainly for electric field sensors, and the response time can be as fast as 0.01 ps～0.1 ps. Their materials are dielectric crystals such as $LiTaO_3$, $LiNbO_3$, CdTe, KTP, and BSO, and their sensitivity is typically on the order of 1 V/m.

13.3　Magnetic measurement

Measurement of magnetic flux by coil　In the measurement of magnetic quantities, magnetic field or magnetic flux is detected on the basis of $B=\mu H$. When the interlinked magnetic flux ϕ is varied temporally, a voltage v is generated by electromagnetic induction and is given by $v=-d\phi/dt$. When ϕ is varied alternately as

磁束の振幅が求められる．測定範囲は 0.001 mWb〜0.1 Wb 程度である．

ホール素子による磁束計測 半導体に発生するホール効果を用いた磁気センサは磁束密度の測定に広く用いられている．また，磁石と組み合わせて，位置センサやモータの回転数センサなどにも用いられる．磁束密度 $\boldsymbol{B}$ の中で，速度ベクトル $\boldsymbol{v}$ で運動する電荷 q の電子は $\boldsymbol{F}_L = q(\boldsymbol{v} \times \boldsymbol{B})$ のローレンツ力を受ける．このとき磁界と垂直に通電のための電界 $\boldsymbol{E}_C$ が作用すると，電子の軌道はサイクロイド曲線を描く．半導体結晶格子と相互作用により，平均的に $\boldsymbol{E}_C$ と角度 θ_H をなす方向に進む．θ_H はホール角と呼ばれる．電子は一方の側面に負電荷として蓄積される．反対側の側面には電荷の不足が生じて，正電位となる．これらの蓄積電荷により生じた電位により，電子の横方向移動は抑制され，平衡状態に達する．このとき電流と直角方向に生じた電位差をホール電圧 V_H と呼び，発生した電圧から $\boldsymbol{B}$ を測定できる．

Fig. 13-12 の座標系において，磁束密度 B_z 中で速度 v_x を持つ電子に対して，ホール電場 E_y が発生する．E_y による力と B_z による力は大きさが釣り合うので，$qE_y=$

$\phi=\phi_0\exp(j\omega t)$, $v=-j\omega\phi_0\exp(j\omega t)$, so the amplitude of ϕ is obtained using v and ω. The measurement range in this method is roughly from 0.001 mWb to 0.1 Wb.

Measurement of magnetic flux by Hall sensor A magnetic sensor using the Hall effect of a semiconductor is widely used in the measurement of flux density. In addition, when combining a Hall sensor with a magnet, sensors for measuring positon and rotation are often utilized. The Lorenz force $\boldsymbol{F}_L$ is generated by the equation $\boldsymbol{F}_L = q(\boldsymbol{v} \times \boldsymbol{B})$ when an electron with charge q is moved at the velocity vector $\boldsymbol{v}$ in the magnetic flux density $\boldsymbol{B}$. When an electric field $\boldsymbol{E}_C$ for current flow is applied in the direction vertical to the magnetic field, the electron orbital path becomes a cycloid curve. Through interaction with a semiconductor crystal lattice, electrons move at the angle of θ_H in the direction of $\boldsymbol{E}_C$ in average motion. The angle θ_H is called the Hall angle. The electrons are accumulated on the sidewalls of crystals as negative charges; on the opposite sidewall, the charge of electrons is in short supply and the wall develops an electric positive potential. By the electric potential generated by the accumulated charges, the lateral motion of electrons is suppressed and reaches an equilibrium state. The electric potential difference vertical to the current is called the Hall voltage V_H, $\boldsymbol{B}$ is measured from the generated voltage.

$qv_x B_z = i_x B_z/N$．ここで，$i_x = Nqv_x$ は電流密度，N はキャリア密度である．I_x は試料を流れる電流である．ホール電圧 V_H は B_z，I_x に比例し，試料の厚さ t に反比例する．比例定数はホール定数 R_H と呼ばれ式(13-3)となる．次に，w を幅として $V_H = E_y w$ であり，$I_x = i_x wt$ であるから，$R_H = E_y/(i_x B_z) = 1/(Nq)$ が得られる．R_H の正負は q の符号による．R_H はキャリア密度 N の逆数に比例する．ホール起電力は N の値が適した半導体において測定しやすい．センサ素子としてヒ化ガリウム(GaAs)，アンチモン化インジウム（InSb）などが用いられる．GaAs ホール素子で，$I_x = 10$ mA で，$B_z = 300$ mT のとき，$V_H = 500$ mV くらいの出力で，$R_H = 150$ V/(AT) 程度の値となる．

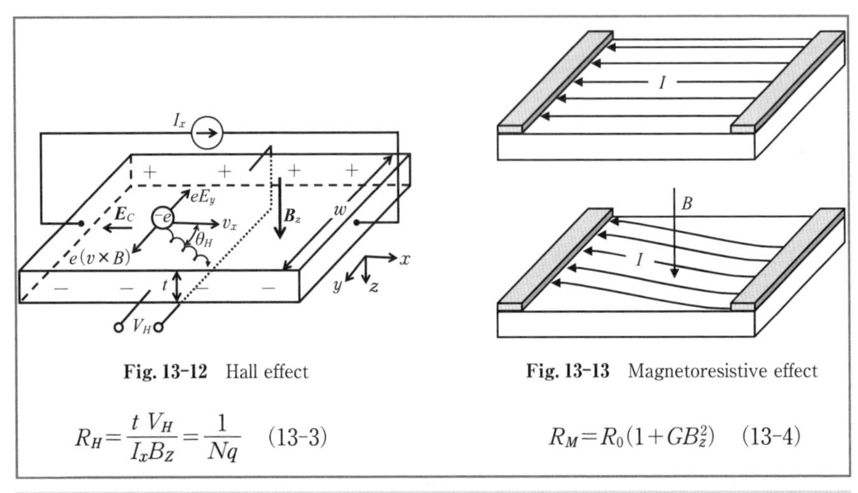

Fig. 13-12　Hall effect

Fig. 13-13　Magnetoresistive effect

$$R_H = \frac{t V_H}{I_x B_Z} = \frac{1}{Nq} \quad (13\text{-}3)$$

$$R_M = R_0(1 + GB_z^2) \quad (13\text{-}4)$$

In the axial coordinates shown in Fig. 13-12, the Hall electric field E_y is generated for the electrons with velocity v_x in the magnetic flux density B_z. As the forces by E_y and B_z are balanced, $qE_y = qv_x B_z = i_x B_z/N$. Here, $i_x = Nqv_x$ is the current density, and N is the carrier density. When I_x is the total current flowing in the sample, Hall voltage V_H is proportional to B_z and I_x, and inversely proportional to the thickness t of sample. The proportional constant in called Hall coefficient R_H, and is given by Eq. (13-3). Next, $V_H = E_y w$ and $I_x = i_x wt$, where w is the width of the sample. The equation $R_H = E_y/(i_x B_z) = 1/(Nq)$ is obtained. The sign of R_H depends on the polarity of q and the value of R_H is inversely proportional to the carrier density N. The Hall voltage is easily measured for semiconductors with an appropriate value of N. Gallium arsenide (GaAs) and indium antimonide (InSb) semiconductors are often used for sensor elements. In the case of GaAs Hall element, a $V_H = 500$ mV is typically obtained under

磁気抵抗効果素子による磁束計測 磁気抵抗効果は，ホール効果が別の現象として現れたものである．Fig. 13-13 に示すように半導体の両端に金属電極を付けて電流を流すとき，磁界の印加によりホール角 θ_H だけ曲げられる．電流を流す電界は電極の金属表面近傍で表面に垂直になるが，電流の流れる距離が全体で増加する．したがって磁界を印加する前の抵抗 R_0 に比較して抵抗 R_M が増加する．R_M の変化より磁束密度 B_z を計測する．R_M と B_z の関係は式(13-4)で与えられる．係数 G はセンサ素子の形状に依存する．

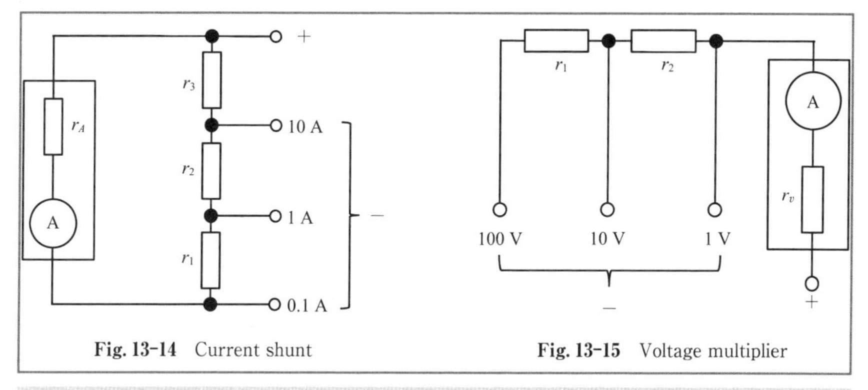

Fig. 13-14 Current shunt

Fig. 13-15 Voltage multiplier

the conditions of I_x=10 mA, and B_z=300 mT, and R_H~150 V/(AT).

Measurement of magnetic flux by magnetoresistive sensor The magnetoresistive effect is a different phenomenon than the Hall effect. As shown in Fig. 13-13, when current flows through the metal electrodes at the ends of a semiconductor, it is bent at the Hall angle θ_H by the application of the magnetic field. The electric field making the current flow into the semiconductor becomes nearly vertical to the metal surface of electrodes, so that the total distance of current flowing is increased. Therefore, the resistance R_M increases compared to the resistance R_0 before applying the magnetic field. From the change in R_M, the magnetic flux density B_z is measured. The relation between R_M and B_z is given by Eq. (13-4). The coefficient G depends on the shape of the sensor element.

【演習問題】

13-1) 電流計の最大測定電流が 100 μA であるとき，Fig. 13-14 に示される端子の電流値が最大測定電流になるためには，分流器の r_1，r_2，r_3 それぞれの抵抗値をどのように決めればよいか．電流計の内部抵抗 r_A を用いて表せ．

13-2) 最大測定電圧が 1 V である電圧計を利用して，Fig. 13-15 の端子に示した値の最大電圧を測定する．倍率器の抵抗 r_1，r_2 を r_v を用いて表せ．

13-3) Fig. 13-8 の自動平衡ブリッジ回路において，$R=10\ \mathrm{k\Omega}$，$\boldsymbol{V}_1=20(1+j)$ のとき出力電圧が $\boldsymbol{V}_2=10(1-j)$ であった．$\boldsymbol{Z}_X$ はどれだけか．また素子は何か．

13-4) キャリア数が $0.4\times10^{17}\ \mathrm{cm}^{-3}$ で厚さ 1 μm のホール素子において，$I_x=10\ \mathrm{mA}$ で出力電圧 $V_H=400\ \mathrm{mV}$ を得た．磁束密度はどれだけか．ただし電子の電荷は 1.6×10^{-19} C とする．

【Problems】

13-1) When the maximum measurable current of an ammeter is 100 μA, express the respective resistances r_1, r_2, and r_3 of the current shunt using the internal resistance r_A of the ammeter so as to make the current values shown beside the terminals in Fig. 13-14 the maximum measurable current values.

13-2) Using a voltmeter with a maximum measurable voltage of 1 V, the maximum voltages shown beside the terminals in Fig. 13-15 are measured. Express the resistances r_1, and r_2 using the internal resistance r_v of the voltmeter.

13-3) In the auto-balancing circuit shown in Fig. 13-8, when $R=10\ \mathrm{k\Omega}$ and $\boldsymbol{V}_1=20(1+j)$, the output voltage is $\boldsymbol{V}_2=10(1-j)$. i) Find $\boldsymbol{Z}_X$; ii) what is the element?

13-4) For a Hall sensor with a carrier density of $0.4\times10^{17}\ \mathrm{cm}^{-3}$ and a thickness of 1 μm, an output voltage $V_H=400\ \mathrm{mV}$ is obtained at $I_x=10\ \mathrm{mA}$. Find the magnetic flux density. Here, the electron charge is 1.6×10^{-19} C.

第14章 計測回路

計測システムにおいて，入力となる計測量はまずセンサによって電気信号に変換されるようになっている．Table 14-1 に本書で取り上げたセンサの出力電気信号を示す．Table 14-1 のように，電気抵抗率や電流などの電位あるいは電位差（電圧）以外の電気信号に変換される場合が多く，それらのセンサ出力信号はそのまま利用しにくいので，さらに電気回路によって電圧信号に変換する必要がある．また，ホール素子や熱電対など計測量を直接電圧信号に変換される場合でも，その信号レベルが微弱

Table 14-1 Electric signals output from sensors

Sensor センサ	Electric output signal 出力電気信号
Strain gauge ひずみゲージ Resistance thermometer 測温抵抗体 Thermistor サーミスタ Thermopile infrared detector サーモパイル Magnetoresistive element 磁気抵抗素子	Electric resistance 電気抵抗
Capacitive displacement sensor 静電容量型変位センサ	Electric capacitance 静電容量
Piezoelectric force sensor 圧電式力センサ Pyroelectric detector 焦電素子	Electric charge 電荷
Photodiode フォトダイオード	Electric current 電流
Hall effect sensor ホール素子 Thermocouple 熱電対	Electric voltage 電圧

Chapter 14 Sensor Signal Conditioning

The input measurand of a physical quantity is converted into an electrical signal by a sensor in the measurement system. Table 14-1 shows the converted electrical signals that have been discussed in this book. The converted electrical signals include electric current and resistance, which are not easy to handle and need to be converted into electric potential or voltage by using electric circuits. Some sensors such as the Hall device or the thermocouple can convert a measurand into electric potential or voltage directly. However, the signal levels are very weak and are associated with noise. It is important to amplify the signal components while reducing the noise

などの理由で，電気回路で増幅したり，さらに不要なノイズ信号を除去したりすることが実用上重要である．本章ではセンサの出力信号に必要な処理を施す計測回路について述べる．

● 14.1　信号を変換する回路

Fig. 14-1 には，センサ電気抵抗値出力 R_s の変化量 ΔR を電圧 v_{out} に変換するのに最もよく利用されているホイートストンブリッジを示す．図に示すように，センサも含めた 4 つの抵抗が点 ABCD 間に四角形状に配置され，点 BD 間の電位差 v_{BD} を回路の電圧出力 v_{out} としている．電源からの電流 i は点 A で i_1，i_2 に分かれる．i_1，i_2 はそれぞれ点 B と点 D を経由して再び点 C で電流 i に再結合し，電源に戻る．説明

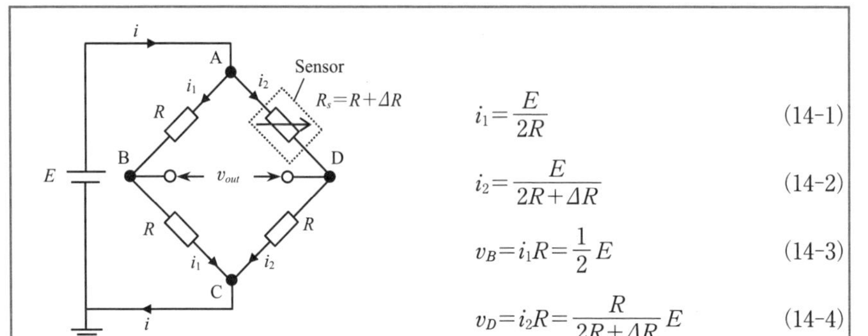

$$i_1=\frac{E}{2R} \tag{14-1}$$

$$i_2=\frac{E}{2R+\Delta R} \tag{14-2}$$

$$v_B=i_1R=\frac{1}{2}E \tag{14-3}$$

$$v_D=i_2R=\frac{R}{2R+\Delta R}E \tag{14-4}$$

Fig. 14-1　Conversion of electric resistance change into voltage output

components for practical use. The signal conditioning circuits used for these purposes are presented in this chapter.

● 14.1　Signal conversion circuits

Fig. 14-1 shows the Wheatstone circuit that converts the change ΔR in the sensor output electric resistance R_s into the voltage v_{out}. Four electric resistors, including the sensor, are connected to the circuit. The potential difference v_{BD} between BD is treated as the output voltage v_{out} of the circuit. The electric current i from the power supply is divided into i_1 and i_2 at point A. i_1 and i_2 are recombined at point C after they have flowed through points B and D respectively. For simplicity, the initial resistances of the four resistors are assumed to be the same R. The output v_{out} corresponds to the change ΔR in the sensor resistance. Point C is connected to the ground, so its potential v_C is zero. Eq. (14-1) can be obtained based on Ohm's law from path ABC. Here, E is

を簡単にするために，4つの抵抗は同じ初期抵抗値 R を持ち，AD 間のセンサ出力抵抗値に変化量 ΔR が生じた場合の出力 v_{out} を求める．また，点 C がグランドに接続されており，その電位 v_C がゼロであるとする．電流 i_1 が流れる経路 ABC においてオームの法則を適用すると，式(14-1)が得られる．ここで E は電源電圧である．同様に，i_2 電流が流れる経路 ADC において式(14-2)が得られる．次に BC 間と CD 間においてオームの法則からそれぞれ電位 v_B と v_D を式(14-3)と(14-4)のように求めることができる．v_{out} は BD 間の電位差であるため，v_{out} と ΔR の関係を式(14-5)のように表すことができ，センサ出力抵抗値の変化量を電圧に変換することができる．

ひずみゲージなどのセンサでは，ΔR が R に比べて十分小さいので，式(14-5)を $\Delta R/R$ の関数としてテイラー展開し，二次以上の非線形項を無視すれば，式(14-6)のように v_{out} と ΔR は近似的に $E/4R$ を感度係数とする線形関係が得られる．v_{out} には

$$v_{out}=v_B-v_D=\left(\frac{1}{2}-\frac{R}{2R+\Delta R}\right)E=\frac{\Delta R}{4R+2\Delta R}E \quad (14\text{-}5)$$

$$v_{out}\approx\frac{E}{4R}\Delta R \quad (14\text{-}6)$$

$$v_{out}\approx\frac{Ek_s}{4d}\Delta d \quad (14\text{-}7)$$

$$v_{in}=v_+-v_- \quad (14\text{-}8)$$

$$v_{out}=v_{in}G \quad (14\text{-}9)$$

$$G\to\infty \quad (14\text{-}10)$$

$$v_{in}\to 0 \quad v_+\approx v_- \quad (14\text{-}11)$$

$$Z_{in}\to\infty \quad (14\text{-}12)$$

$$i_{in}=\frac{v_{in}}{Z_{in}}\to 0 \quad (14\text{-}13)$$

$$Z_{out}=0 \quad (14\text{-}14)$$

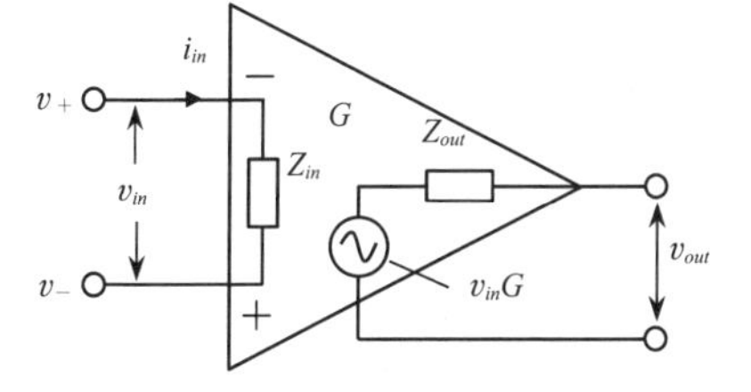

Fig. 14-2　Schematic of an operational amplifier

the voltage of the power supply. Similarly, Eq. (14-2) can be obtained from path ADC. The potential v_B and v_D are evaluated as Eqs. (14-3) and (14-4) by applying Ohm's law between BC and CD respectively. Since v_{out} is the potential difference between BD, the relationship between v_{out} and ΔR is expressed in Eq. (14-5), from which the change in the sensor resistance can be converted into voltage.

For sensors such as strain gauges, ΔR is much smaller than R. Eq. (14-5) is thus approximated by the Taylor series in Eq. (14-6), where second or higher order nonlinear terms are omitted. As Eq. (14-6) shows, v_{out} is a linear function of ΔR with the sensitivity coefficient $E/4R$. Because there is no offset term in v_{out}, it is possible to amplify v_{out} by using an amplifier circuit with a high amplification. When a bridge

オフセット電圧が存在しないため，後続の増幅回路は大きな増幅倍率が設定できる利点がある．ブリッジ回路を Fig. 6-9 のひずみゲージに応用した場合，式(6-25)を式(14-6)に代入して v_{out} と変形量 Δd の関係式（式(14-7)）を得ることができる．

静電容量，電流，電荷など電気抵抗以外のセンサ出力を電圧に変換するには，Fig. 14-2 に示す演算増幅器（オペアンプ）がよく利用される．式(14-8)から(14-14)までにオペアンプの特性を示す．Fig. 14-2 のように，オペアンプには電位が v_+ の非反転入力端子（＋）と電位が v_- の反転入力端子（－）があり，その電位差である v_{in}（式(14-8)）が増幅されて出力端子から電圧 v_{out} が出力される．式(14-9)の G はオペアンプの開ループ電圧利得（増幅率）である．理想的なオペアンプでは，G は無限大のため（式(14-10)），v_{out} と G の比である v_{in} はほぼゼロとなり，入力端子の電位 v_+，v_- はほぼ同じになる（式(14-11)）．また入力インピーダンス Z_{in} は無限大のため（式(14-12)），オペアンプの入力電流 i_{in} はほぼゼロとなる（式(14-13)）．さらに，出力インピーダンスはゼロであり（式(14-14)），後続回路の入力インピーダンスに影響されずに式(14-9)は常に成立する．

オペアンプを利用してセンサ静電容量出力 C_s の変化量 ΔC を電圧 v_{out} に変換する回路を Fig. 14-3 に示す．電源から角周波数 ω の交流電圧 E が出力される．コンデン

circuit is employed for the strain gauge shown in Fig. 6-9, the relationship between v_{out} and the measured deformation Δd can be written in Eq. (14-7) by substituting Eq. (6-25) into Eq. (14-6).

An operational amplifier, as shown in Fig. 14-2, is often utilized for conversion of sensor outputs of electric capacitance, current, and charge. The amplifier characteristics are shown in Eqs. (14-8) to (14-14). An operational amplifier has a non-inverting input (+) with a potential v_+ and an inverting input (−) with a potential v_-. The voltage v_{in} in Eq. (14-8) between the two inputs is amplified to v_{out}. The G in Eq. (14-9) is the open-loop gain and is infinite for an ideal operational amplifier, as shown in Eq. (14-10). v_{in}, which is the ratio of v_{out} to G, is thus approximately zero and v_+ is approximately equal to v_-, as shown in Eq. (14-11). Because the input impedance Z_{in} in Eq. (14-12) is infinite, the input current i_{in} in Eq. (14-12) is approximately zero. Eq. (14-9) is always satisfied regardless of the input impedance of the following circuit because the amplifier output impedance in Eq. (14-14) is zero.

An operational amplifier circuit for converting the change ΔC in a sensor output capacitance C_s into the voltage v_{out} is shown in Fig. 14-3. An AC voltage E with an

サ C_1，C_s のインピーダンス Z_1，Z_s はそれぞれ式(14-15)と(14-16)のようになる．点 A，B の電位は同じとみなせるので，コンデンサ C_1 の印加電圧 v_1 は E と同じとなり（式(14-17)），電流 i は式(14-18)のように表せる．同じ電流 i がセンサである C_s を流れるので，C_s の印加電圧 v_s は式(14-19)のように求められる．なお，出力電圧 v_{out} は v_s と同じである（式(14-20)）．この回路を Fig. 6-2 の静電容量型変位計に応用した場合は，式(6-6)を式(14-19)に代入して v_{out} と変位量 Δd_z の関係式を式(14-21)のように得ることができる．式(6-7)の ΔC と Δd，また式(14-19)のように v_s (v_{out}) と ΔC はそれぞれ非線形の関係であるが，両者を組み合わせることによって v_{out} と Δd_z は理想的な線形関係になる．

Fig. 14-4 にはセンサ電荷出力 Q_s を電圧 v_{out} に変換する回路を示す．チャージアンプと呼ばれるこの回路では，センサに発生した電荷 Q_s は電流によってそれぞれセン

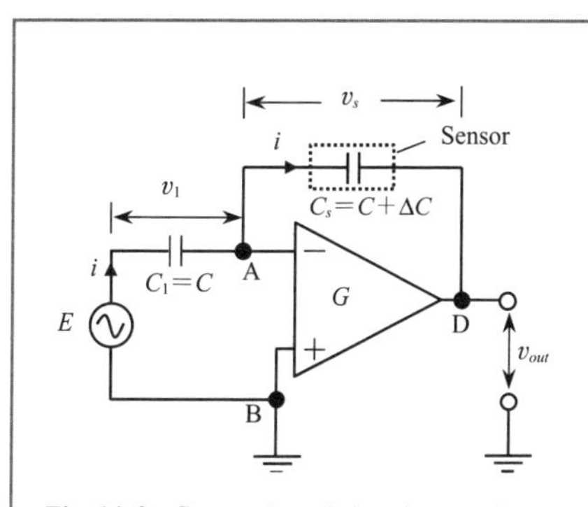

Fig. 14-3　Conversion of electric capacitance change into voltage output

$$Z_1=\frac{1}{j\omega C_1}=\frac{1}{j\omega C} \tag{14-15}$$

$$Z_S=\frac{1}{j\omega C_S}=\frac{1}{j\omega(C+\Delta C)} \tag{14-16}$$

$$v_1=E \tag{14-17}$$

$$i=\frac{v_1}{Z_1}=j\omega CE \tag{14-18}$$

$$v_S=Z_2 i=\frac{C}{C+\Delta C}E \tag{14-19}$$

$$v_{out}=v_S \tag{14-20}$$

$$v_{out}=\left(1+\frac{\Delta d_z}{d_z}\right)E \tag{14-21}$$

angular frequency ω is output from the power supply. The impedances Z_1 and Z_s of capacitors C_1 and C_s are shown in Eqs. (14-15) and (14-16) respectively. The voltage v_1 in Eq. (14-17) over C_1 is approximately E, while the current i flowing through C_1 is obtained in Eq. (14-18). Eq. (14-19) shows the voltage v_s over the sensor C_s, through which the same i flows. The output voltage v_{out} in Eq. (14-20) is the same as v_s. When the circuit is applied to the capacitive displacement sensor in Fig. 6-2, a linear relationship between v_{out} and displacement Δd_z can be obtained in Eq. (14-21) by substituting Eq. (6-6) into Eq. (14-19).

Fig. 14-4 shows the circuit, called a charge amplifier, for converting the sensor charge output Q_s into the voltage v_{out}. Q_s is moved by the current to the internal capacitor C_1 of the sensor and the feedback capacitor C_f. Q_s in Eq. (14-22) is the sum

サの内部容量 C_1 とオペアンプのフィードバック容量 C_f に移動する．C_1 と C_f に蓄えられる電荷をそれぞれ Q_1，Q_f とすると，Q_s は Q_1 と Q_f の和となる（式(14-22)）．一方，C_1 にかかる電圧 v_1 はオペアンプの入力電圧となるので，式(14-11)より，v_1 はほぼゼロとなり（式(14-23)），v_1 と C_1 の積である Q_1 もゼロとみなせる（式(14-24)）．そのため，センサ電荷出力 Q_s は電流 i によってほぼ全部 C_f に移動し（式(14-25)），C_f にかかる電圧 v_f およびそれと等しいチャージアンプの電圧出力 v_{out} は式(14-26)のように求められる．この回路を第8章の圧電型力センサに応用した場合は，式(8-15)と式(14-26)から v_{out} と力の関係式を式(14-27)のように示すことができる．

一方，実際のオペアンプには Fig. 14-4 に示すバイアス直流電流 I_B が存在する．こ

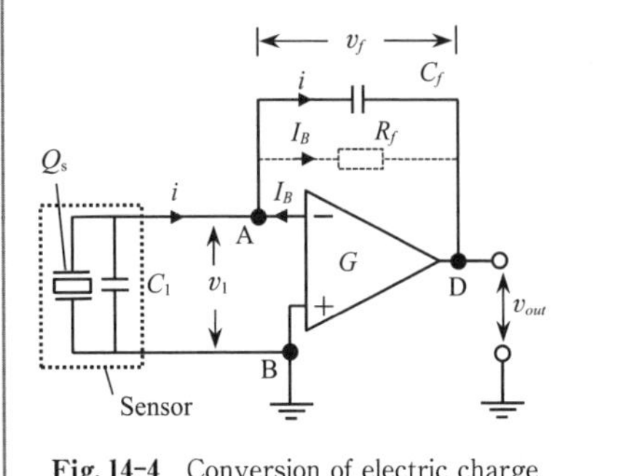

Fig. 14-4 Conversion of electric charge change into voltage output

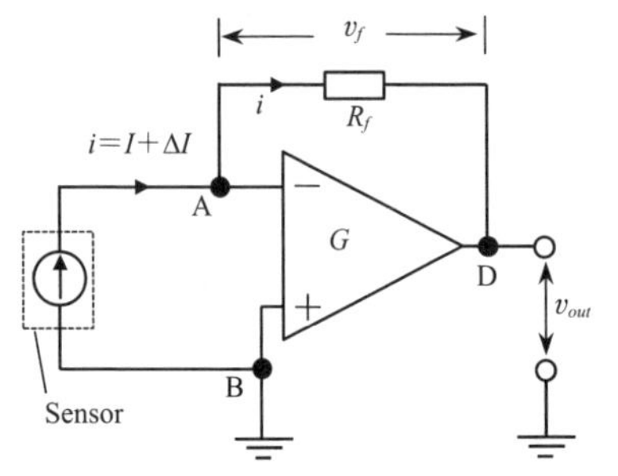

Fig. 14-5 Conversion of electric current change into voltage output

$$Q_s = Q_1 + Q_f \quad (14\text{-}22)$$

$$v_1 \approx 0 \quad (14\text{-}23)$$

$$Q_1 = v_1 C_1 \approx 0 \quad (14\text{-}24)$$

$$Q_f = Q_s - Q_1 \approx Q_s \quad (14\text{-}25)$$

$$v_{out} = v_f = -\frac{Q_f}{C_f} = -\frac{Q_s}{C_f} \quad (14\text{-}26)$$

$$v_{out} = -\frac{d_{33}}{C_f} F \quad (14\text{-}27)$$

$$v_{out} = v_f = -iR_f = -(I + \Delta I) R_f \quad (14\text{-}28)$$

of Q_1 and Q_f, which are the charges in C_1 and C_f respectively. The voltage v_1 in Eq. (14-23) over C_1, which is the input voltage of the operational amplifier, is approximately zero, based on Eq. (14-11). Q_1 in Eq. (14-24), the product of v_1 and C_1, is also zero. Therefore, almost all of Q_s is moved to C_f, as shown in Eq. (14-25). The voltage v_f over C_f and the charge amplifier output voltage v_{out} are thus evaluated in Eq. (14-26). When the circuit is applied to the piezoelectric force sensor, the relationship between v_{out} and the force can be obtained in Eq. (14-27) by substituting Eq. (8-15) of Chapter 8 into Eq. (14-26).

On the other hand, there exists a bias current I_B in an actual operational amplifier, as shown in Fig. 14-4. Although I_B is very low, since C_f is continuously charged by I_B,

の電流は微小であるが，C_f を充電し続けるため，やがて v_f と v_{out} はオペアンプの電源電圧の近くまで達してしまい，飽和してしまう問題がある．そこで，図の点線で示すように，オペアンプにフィードバック抵抗 R_f を追加する．コンデンサの直流インピーダンスが無限大のため，I_B を抵抗 R_f を介して逃がすことができる．v_{out} には $-I_B R_f$ というオフセット成分が生じるが，Q_s の交流成分を I_B に影響されずに検出できる．

Fig. 14-5 にはセンサ電流出力 I の変化量 ΔI を電圧 v_{out} に変換する I-V 変換回路を示す．センサからの出力電流 i は抵抗 R_f を流れるため，R_f にかかる電圧 v_f と等しい出力電圧 v_{out} は式(14-28)のように求めることができる．

● 14.2 信号を増幅する回路

信号変換回路やセンサの電圧出力信号を拡大したい場合は，Fig. 14-6 の差動増幅回路がよく利用される．点 A，B の電位 v_A，v_B が式(14-29)のように定められるので，

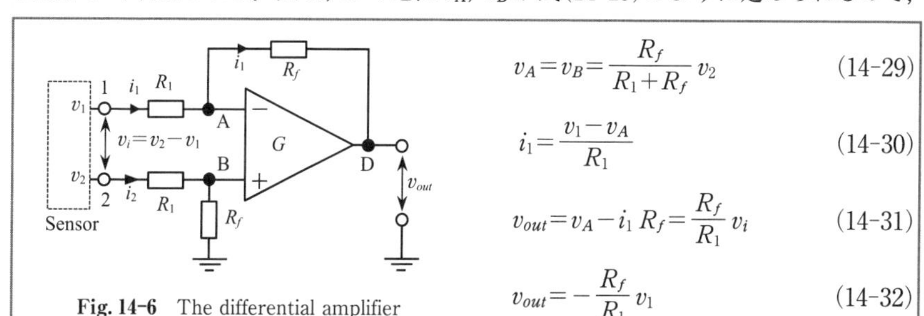

Fig. 14-6 The differential amplifier

$$v_A = v_B = \frac{R_f}{R_1 + R_f} v_2 \quad (14\text{-}29)$$

$$i_1 = \frac{v_1 - v_A}{R_1} \quad (14\text{-}30)$$

$$v_{out} = v_A - i_1 R_f = \frac{R_f}{R_1} v_i \quad (14\text{-}31)$$

$$v_{out} = -\frac{R_f}{R_1} v_1 \quad (14\text{-}32)$$

v_f and v_{out} will be saturated to the voltage of the power supply of the operational amplifier. A feedback resistance R_f is added in Fig. 14-4 to solve this problem. Because the DC impedance of a capacitor is infinite, I_B will flow through R_f instead of C_f. Although there exists an offset $-I_B R_f$ in v_{out}, the change in Q_s, which is an AC component, can be detected without any influence by I_B.

Fig. 14-5 shows the I-V circuit for converting the sensor current output i into the voltage v_{out}, which equals the voltage v_f over R_f, through which i flows, as shown in Eq. (14-28).

● 14.2 Signal amplification circuits

The differential amplifier in Fig. 14-6 is often used for amplification of the voltage output of a sensor or a signal conversion circuit. The current i_1 can be evaluated in Eq. (14-30) by the potentials v_A and v_B at A and B, as shown in Eq. (14-29). The difference v_i between v_2 and v_1 is amplified to v_{out} in Eq. (14-31). The ratio of the

電流 i_1 は式(14-30)のように計算でき，v_2 と v_1 の差である増幅回路の入力電圧 v_i を式(14-31)のように v_{out} まで増幅できる．抵抗値 R_f と R_1 の比は回路の増幅率となる．なお，v_1 と v_2 のどちらかがゼロ電位になる場合でも，v_{out} と v_i の関係式は式(14-31)のとおりである．v_2 の電位がゼロのときは，式(14-32)のように，v_1 と v_{out} の符号が逆となるので，反転増幅器と呼ばれる．

一方，Fig. 14-6 の回路では R_1 と R_f の和が回路の入力インピーダンスとなる．Fig. 14-7 に示すように，増幅回路の入力インピーダンス Z_i がセンサの負荷となり，増幅回路の入力電圧 v_i とセンサの電圧出力 v_s の関係が式(14-33)に表せる．センサの出力インピーダンス Z_{s_o} は通常ゼロではないので，Z_i が無限大という理想的な場合以外では，v_s の一部しか増幅回路に入力されないことが分かる．センサの出力電圧は通常微弱なので，Z_i をなるべく大きくして v_s の損失は避ける必要がある．そのために

$$v_i = \frac{Z_i}{Z_i + Z_{s\_o}} v_s \quad (14\text{-}33)$$

$$v_{out} = \left(1 + \frac{2R_2}{R_1}\right) v_i \quad (14\text{-}34)$$

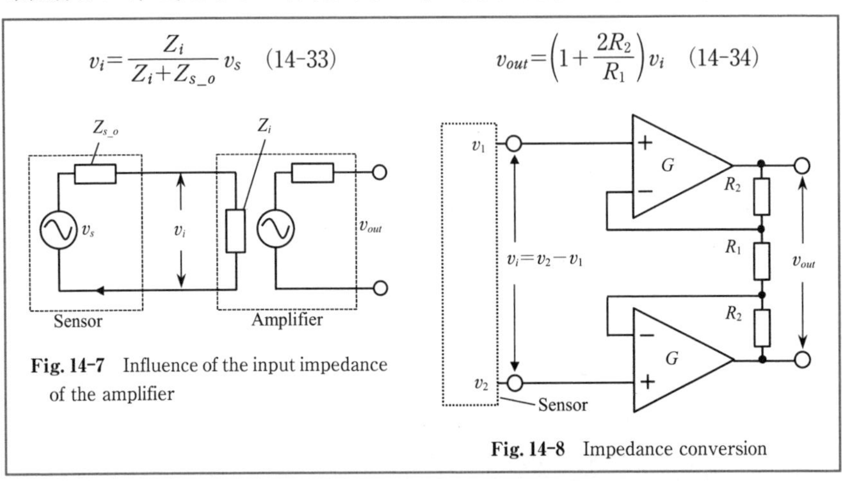

Fig. 14-7　Influence of the input impedance of the amplifier

Fig. 14-8　Impedance conversion

resistances R_f to R_1 is the amplification factor of the circuit. The relationship between v_{out} and v_i in Eq. (14-31) is also satisfied when v_1 or v_2 is zero. v_1 and v_{out} in Eq. (14-32) have opposite signs when v_2 is zero, in that case, the circuit is called an inverting amplifier.

The sum of R_1 and R_f is the input impedance of the circuit in Fig. 14-6. The input impedance Z_i of the amplifier is the load of the sensor, as shown in Fig. 14-7, and the relationship between the amplifier input voltage v_i and the sensor output voltage v_s is expressed by Eq. (14-33). Because the output impedance Z_{s_o} of the sensor is not zero, only part of v_s can be input into the amplifier for amplification if Z_i is not infinite. Therefore a large Z_i is necessary to reduce the loss of v_s. Fig. 14-8 shows the circuit used for this purpose. The input impedance of the circuit can be improved to equal

用いられる回路を Fig. 14-8 に示す．入力インピーダンスをオペアンプの入力インピーダンスまで高められるのがこの回路の特徴である．その後段に Fig. 14-6 の差動増幅回路を接続した回路（インスツルメンテーションアンプと呼ばれる）は計測用増幅回路としてよく利用される．Fig. 14-8 の回路において v_{out} と v_i の関係式は式(14-34)のとおりである．なお，インピーダンス変換用回路として，ボルテージフォロアもよく用いられる．

● 14.3　信号をきれいにする回路

センサの出力には，計測量に直接関連する信号成分以外にもノイズと呼ばれる不要な成分が含まれる．信号成分とノイズ成分の周波数分布の違いを利用して，センサ出力からノイズ成分を除去し，信号成分だけを残すフィルタ回路と，特定の周波数の信

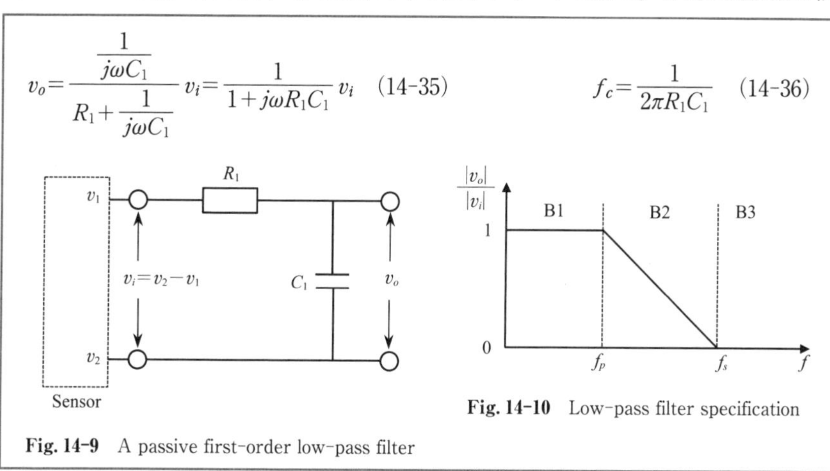

$$v_o=\frac{\frac{1}{j\omega C_1}}{R_1+\frac{1}{j\omega C_1}}v_i=\frac{1}{1+j\omega R_1C_1}v_i \quad (14\text{-}35)$$

$$f_c=\frac{1}{2\pi R_1C_1} \quad (14\text{-}36)$$

Fig. 14-9　A passive first-order low-pass filter

Fig. 14-10　Low-pass filter specification

that of the operational amplifier. The differential amplifier in Fig. 14-6 can be connected after this circuit to generate a circuit called the instrumentation amplifier, which is often used in measurement and instrumentation. The relationship between v_{out} and v_i in Fig. 14-8 is shown in Eq. (14-34). The impedances of a circuit can also be converted by using a voltage follower circuit.

● 14.3　Signal filtering circuits

Electronic filters can be employed to remove or reduce the noise components so that only the signal components remain in the sensor output by making use of the difference between the frequencies of the noise components and the signal components. A lock-in amplifier can be employed to take the signal component with a

号成分のみをセンサ出力から取り出すロックインアンプがある．

抵抗とコンデンサだけで構成されるフィルタ回路はパッシブフィルタと呼ばれる．Fig. 14-9 に入力信号 v_i に存在する高周波ノイズ成分をカットし，低周波の信号成分を通過させる一次パッシブローパスフィルタ回路を示す．その出力 v_o と遮断周波数 f_c はそれぞれ式(14-35)，(14-36)のとおりである．一方，ローパスフィルタの振幅特性を模式的に Fig. 14-10 に示すことができる．周波数 f_p 以下の周波数範囲 B1 は通過域，f_s 以上の範囲 B3 は阻止域，その間の範囲 B2 は遷移域と呼ばれる．理想的なローパスフィルタでは B2 は存在せず，B1 と B3 にはそれぞれ信号成分とノイズ成分しか含まれないことを条件に，ノイズ成分をカットし，信号成分だけを出力するようになっている．したがって，実際の場合でも，B2 の範囲を極力狭くする必要がある．そのためには，Fig. 14-9 の一次フィルタを直列させて高次のローパスフィルタ

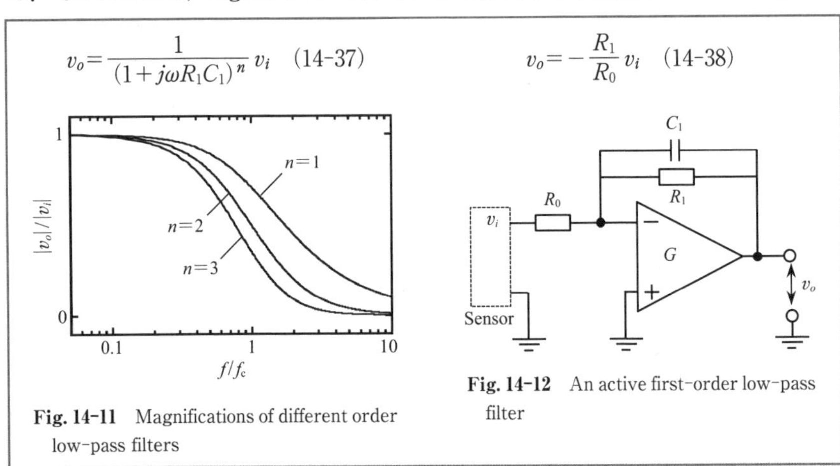

$$v_o = \frac{1}{(1+j\omega R_1 C_1)^n} v_i \quad (14\text{-}37)$$

$$v_o = -\frac{R_1}{R_0} v_i \quad (14\text{-}38)$$

Fig. 14-11　Magnifications of different order low-pass filters

Fig. 14-12　An active first-order low-pass filter

specific frequency out of the sensor output.

A passive filter can be constructed with resistors and capacitors. Fig. 14-9 shows a passive first order, low-pass filter that cuts off high-frequency noises and passes through low-frequency signal components in the input v_i. The output v_o and the cutoff frequency f_c are shown in Eqs. (14-35) and (14-36) respectively. Fig. 14-10 shows a schematic of the amplitude characteristics of a low-pass filter. The frequency ranges B1 below f_p, B3 above f_s, and B2 between f_p and f_s are called the pass band, the stop band, and the transition band respectively. An ideal low-pass filter works on the assumption that there are only signals in B1 and noises in B3, without the existence of B2. In practical cases, it is also necessary to shorten the range of B2. For this purpose, multiple first order filters in Fig. 14-9 are connected in series to construct a higher

回路を構成する．式(14-37)にn次ローパスフィルタの出力を示し，次数によって出力の振幅特性の変化をFig. 14-11に示す．図から分かるように，高次になるにつれて，遷移域B2は狭くなり，フィルタ効果は向上する．さらに，Fig. 14-12に示すように，オペアンプを活用して，フィルタと増幅回路を1つの回路で実現するアクティブフィルタもある．反転増幅器を利用した一次ローパスフィルタの図の例では遮断周波数f_cは式(14-35)に示されるパッシブのものと同じであるが，式(14-38)のように信号を増幅することができる．

一次パッシブハイパスフィルタをFig. 14-13に示す．低周波ノイズをカットし，高周波信号を通すこの回路の入出力関係およびカットオフ周波数を式(14-39)と(14-40)に示す．また，ローパスとハイパスフィルタを組み合わせることによって，一部だけ

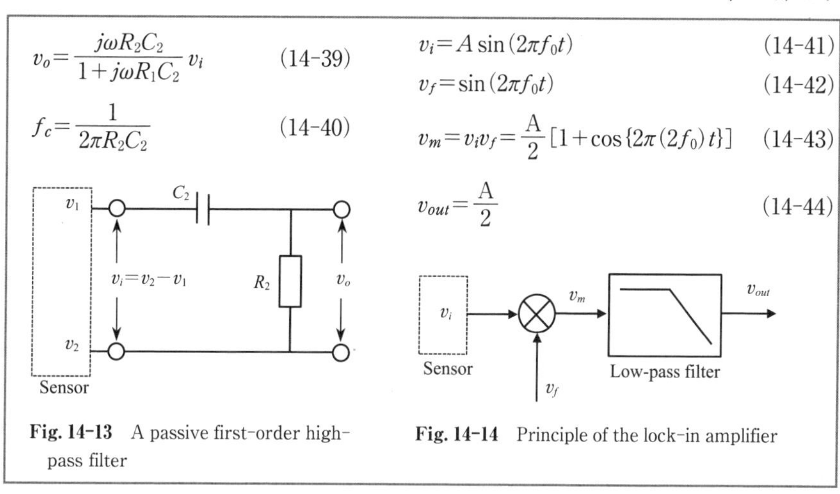

Fig. 14-13　A passive first-order high-pass filter

Fig. 14-14　Principle of the lock-in amplifier

order low-pass filter. The output of an nth order low-pass filter and its amplitude characteristics are shown in Eq. (14-37) and Fig. 14-11 respectively. B2 becomes narrower when the order n of the filter increases and the filtering effect is improved. Fig. 14-12 shows an active filter with filtering and amplification functions when an operational amplifier is employed. The cutoff frequency of this active first order low-pass filter with an inverting amplifier is the same as that of the passive first order low-pass filter in Eq. (14-35), but the gain in the pass band increases, as shown in Eq. (14-38).

Fig. 14-13 shows that a passive first order high-pass filter cuts off low-frequency noises and passes through high-frequency components. The output and the cutoff frequency are shown in Eqs. (14-39) and (14-40). The signals over a certain range of frequencies can be passed through by using a band-pass filter, which is a combination

の周波数範囲の信号を通すバンドパスあるいはノイズをカットするバンドストップフィルタが実現できる．また，これらの種類のアクティブフィルタも同様にある．

Fig. 14-14 にはある特定の周波数 f_0 を持つセンサ出力の信号成分 v_i（式(14-41)）を周りのノイズ成分から取り出し，信号成分の振幅 A を求めるロックインアンプの原理図を示す．信号成分と同じ周波数 f_0 を持つ参照信号 v_f（式(14-42)）を v_i と掛け合わせることによって，式(14-43)の信号 v_m を得ることができる．v_m には大きさが $A/2$ の直流成分と f_0 の二次高調波の項が含まれるが，v_m をローパスフィルタに通して得られる v_{out}（式(14-44)）には，二次の高調波が除去され，v_i の振幅 A だけが残る．

【演習問題】

14-1) Fig. 8-4 に示すロードセルで同様なひずみゲージを4枚使う差動構造をとっている．それに利用されるブリッジ回路を構成せよ．また，ブリッジ回路の電圧回路 v_{out} の出力とひずみゲージの抵抗変化量 ΔR の関係式を導け．

14-2) Fig. 14-3 において v_{out} が 0.8 V として測定されたときの ΔC を求めよ．ただ

of a low-pass filter and a high-pass filter. There are also active types for such filters.

Fig. 14-14 shows the principle of a lock-in amplifier. The amplitude A of the signal component v_i in Eq. (14-41) with a certain frequency f_0 is evaluated by extracting the signal from the sensor output without any influence from noises containing different frequencies. Multiplying v_i by a reference signal v_f in Eq. (14-42) with the frequency of f_0 gives v_m in Eq. (14-43). v_m consists of a DC component $A/2$ and a second-order harmonic component of f_0. The latter is cut off and the former remains in v_{out} of Eq. (14-44) by passing v_m through a low-pass filter. A can thus be obtained from v_{out}.

【Problems】

14-1) A differential structure with two identical strain gauges is employed in the load cell shown in Fig. 8-4. Design a Wheatstone bridge circuit for it, and derive the relationship between the output voltage of the circuit and the changes ΔR in the strain gauges.

14-2) Evaluate the ΔC in Fig. 14-3 when v_{out} is measured at 0.8 V. E and C are 1 V and 1.2 μF respectively.

し，E と C はそれぞれ 1 V と 1.2 μF とする．

14-3) Fig. 14-4 の回路を水晶型力センサに応用する場合を考える．水晶の d_{33} を調べたうえで，1 kN の力 F に対応する v_{out} が -0.2 V になるようにするための C_f を決めよ．

14-4) Fig. 14-9 の回路において，周波数 f が $2f_c$ のときの出力 v_o を示せ．

14-5) ロックインアップの時定数 τ からローパスフィルタの遮断周波数 f_c を設定することができる．f_c を τ で表せ．

14-3) Consider the circuit in Fig. 14-4 when it is applied to a quartz force sensor. Investigate d_{33} of quartz and calculate the C_f when v_{out} becomes -0.2 V for a 1 kN force F.

14-4) Determine the v_o in Fig. 14-9 when f is equal to $2f_c$.

14-5) The cutoff frequency f_c can be determined from the time constant τ for a lock-in amplifier. Describe the relationship between f_c and τ.

参考文献
References

1. JIS Z 8103：2000 計測用語
2. JIS Z 2241：2011 金属材料引張試験方法
3. JIS Z 2242：2005 金属材料のシャルピー衝撃試験方法
4. JIS Z 2243：2008 ブリネル硬さ試験―試験方法
5. JIS Z 2244：2009 ビッカース硬さ試験―試験方法
6. JIS Z 2245：2011 ロックウェル硬さ試験―試験方法
7. JIS Z 2246：2000 ショア硬さ試験―試験方法
8. JIS Z 2251：2009 ヌープ硬さ試験―試験方法
9. JIS Z 6253-3：2012 加硫ゴム及び熱可塑性ゴム―硬さの求め方―第3部：デュロメータ硬さ
10. 産業技術総合研究所 計量標準総合センター訳・監修，『国際単位系(SI)日本語版』，2006
11. 富士セラミックス，『圧電型加速度ピックアップテクニカル・ハンドブック』
12. 富士セラミックス，『圧電セラミックステクニカル・ハンドブック』
13. FDK，『技術資料 圧電セラミックス』
14. エー・アンド・デイ，『ロードセル入門』
15. 宮下文秀，質量，容量の正確な計量，ぶんせき，**1**，1-10，2008
16. 杉浦宣裕，『天びんの構造と原理』，ザルトリウス
17. 三谷幸寛，ヤング率の測定技術について，IIC Review，**43**，30-34，2010
18. テクロック，『技術資料 デュロメータ/IRHD 硬さ計』
19. 今井秀孝，『測定不確かさ評価の最前線』，日本規格協会，2013
20. 浜松ホトニクス，『技術資料 光電子増倍管と関連製品』
21. 浜松ホトニクス，『技術資料 Si フォトダイオード』
22. 南　茂夫，木村一郎，荒木　勉，『はじめての計測工学 改訂第2版』，講談社，2012
23. 西原主計，山藤和男，松田康広，『計測システム工学の基礎（第3版）』，森北出版，2012
24. 土屋喜一，『大学課程 計測工学（第3版）』，オーム社，2000
25. 前田良昭，木村一郎，押田至啓，『計測工学』，コロナ社，2001

26. 萩原将文，『ディジタル信号処理（第2版）』，森北出版，2014
27. 上田政文，『湿度と蒸発』，コロナ社，2000
28. 南任靖雄，『センサと基礎技術』，工学図書，1994
29. 清野次郎，近藤昭治，『センサ工学入門』，森北出版，1988
30. 塩山忠義，『センサの原理と応用』，森北出版，2002
31. John R. Taylor（林　茂雄，馬場　凉訳），『計測における誤差解析入門』，東京化学同人，2000
32. 中村邦雄編著，石垣武夫，冨井　薫，『計測工学入門（第3版）』，森北出版，2015
33. 下田　茂，穂苅　久，愛原惇士郎，高野英資，長谷川富市，『計測工学』，コロナ社，1982
34. BIPM, "*The International System of Units (SI), 8th Edition*", 2006
35. JCGM, "*JCGM 100：2008 Evaluation of measurement data —— Guide to the expression of uncertainty in measurement (GUM)*", 2008
36. JCGM, "*JCGM 200：2012 International Vocabulary of Metrology —— Basic and General Concepts and Associated Terms (VIM 3rd edition)*", 2012
37. Richard S. Figliola, Donald E. Beasley, "*Theory and Design for Mechanical Measurements*", John Wiley & Sons, 2015
38. Stephanie Bell, "*A Beginner's Guide to Uncertainty of Measurement*", National Physical Laboratory, 2001
39. Michael F. Ashby, David R. H. Jones, "*Engineering Materials 1*", Butterworth-Heinemann, 1996
40. Lion Precision, "*Understanding Sensor Resolution Specifications and Effects on Performance*", General Sensor TechNote LT05-0010, 2014
41. ADE MicroSense, http://www.microsense.net
42. Wei Gao, "*Precision Nanometrology*", Springer, 2010
43. 臼田　孝，『新しい1キログラムの測り方　科学が進めば単位が変わる』，講談社，2018
44. https://ja.wikipedia.org/wiki/新しいSIの定義
45. Wei Gao, "*Surface Metrology for Micro- and Nanofabrication*", Elsevier, 2020

演習問題解答
Answers to Selected Problems

1-1) LT^{-2}, L^2MT^{-2}, T^{-1}　**1-2)** 1.346×10^{10}　**1-3)** kg　**1-4)** ローマン体（Roman type）　**1-5)** メートル原器（The international prototype meter bar）

2-1) 精度（accuracy）　**2-2)** $y(t)=\dfrac{1}{a_0}\left(1-e^{-\frac{t}{\tau}}\right)$, $\tau=\dfrac{a_1}{a_0}$, $T_d=-\dfrac{a_1}{a_0}\ln 0.5$　**2-3)** $\sigma=0.12$　**2-4)** 再現性（reproducibility）　**2-5)** 6

3-1) $u(Rpt)=0.577\,\mu\text{m}/\sqrt{16}=0.144\,\mu\text{m}$　**3-2)** アッベの誤差，分解能など（Abbe error, Measurement resolution etc.）　**3-3)** GUM 参照.（Refer to GUM）

4-1) コンデンサを 4 つ追加し，その比を上位から 128：64：32：16：8：4：2：1：1 に設計する.（Add four capacitors to Fig. 4-2 and set the capacitor ratios to be 128：64：32：16：8：4：2：1：1, from high to low.）

4-2) $(1/\pi f)\sin(2\pi f)$

4-3)

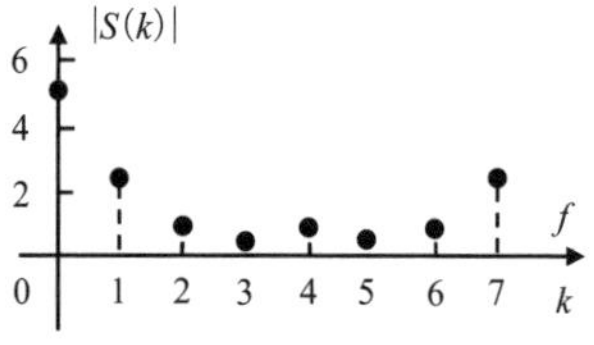

$S(k)$ DFT of $s(n)$

$$S(k)=e^{-j\frac{\pi}{2}k}\frac{\sin\left(\frac{5\pi}{8}k\right)}{\sin\left(\frac{\pi}{8}k\right)}$$

$S(0)=5$, $S(1)=-j2.414$, $S(2)=1$, $S(3)=-j0.414$, $S(4)=1$, $S(5)=j0.414$, $S(6)=1$, $S(7)=j2.414$

4-4) $S(0)=0$, $S(1)=0$, $S(2)=4$, $S(3)=0$, $S(4)=0$, $S(5)=0$, $S(6)=4$, $S(7)=0$

5-1) 0.60 ± 0.05 N/mm　**5-2)** (1) 13 ± 6 mm, (2) 21.0 ± 1.5 s, (3) -760 ± 30 MPa
5-3) $P_{60}(|0.4|\geqq r_0)=0.2\%$

6-1) $\varDelta C=0.278$ pF　**6-2)** $p=1.2f$, $q=6f$　**6-3)** $\beta=45°$　**6-4)** $\nu_s=0.34$, $k_s=1.68$　**6-5)** はりの固定端（the fixed end of the beam）

7-1) λ　**7-2)** 20.98 kHz　**7-3)** $u_v=\left(\dfrac{\tan\alpha}{2\cos\alpha}\dfrac{V}{f_s}\varDelta f\right)u_\alpha$

7-4) $G_x(j\omega)=\dfrac{X_r(j\omega)}{X(j\omega)}=\dfrac{\omega^2}{-\omega^2+j2\xi\omega\omega_n+\omega_n^2}$ 7-5) $\omega=1.272\omega_n$

8-1) 1×10^{-5} 8-2) $F=\dfrac{2Ebh^2}{3Lk_sR}\Delta R$ 8-3) $C_d=\varepsilon_{33}\dfrac{S}{h}$, $R=\rho\dfrac{h}{S}$, $RC_d=\rho\varepsilon_{33}$

8-4) 0.07 N 8-5) 0.15 g

9-1) E=200 GPa. 2本の吊り線部以外は試験片に対し接触する部分がないため精度の高い測定が可能となる.（This method has the potential for highly accurate measurements because there is nothing in contact with the plate except the two suspension lines.）

9-2) HV=210

9-3) ロックウェル硬度（Rockwell hardness testing method）. 測定者自身が初期に基準値としてのゼロ点を実験的に設定するため, 誤差の少ない測定が可能となる.（Because a datum position is set firstly by operator.）

9-4) 軟質材料の硬さは, 弾性変形に対する抵抗値を意味し, 硬質材料の硬さは, 塑性変形の程度を意味する.（Hardness of the soft material is recognized as a physical quantity representing the degree of elastic deformation. One of the hard material is recognized degree of plastic deformation.）

10-1) (1) 15 m/s, (2) 29 m/s, (3) 111 m/s 10-2) (1) 0.52 kPa, (2) 3.1 kPa

10-3) (1) 1.6 kPa, (2) 125 kPa, (3) 2.0 kPa 10-4) (1) 1.5×10^{-5} m^2/s, (2) 1.0×10^{-6} m^2/s

11-1) 300.15 K

11-2)

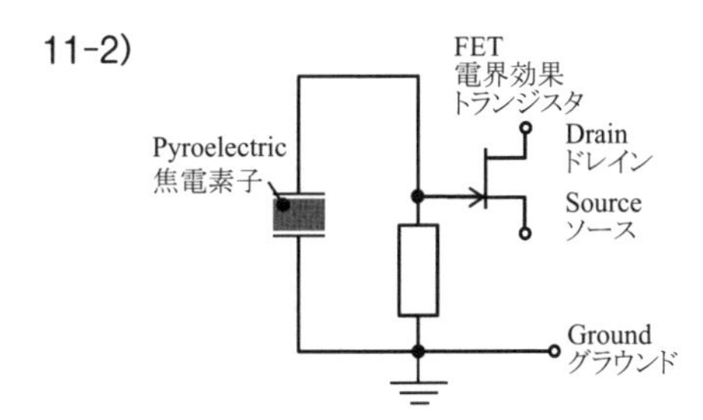

11-3) 標準大気圧 P_0=1013 hPa として, 相対湿度 63.1%, 絶対湿度 14.53 g/m^3（Relative humidity : 63.1%, Absolute humidity : 14.53 g/m^3 under the condition of

standard atmospheric pressure of 1013 hPa)

11-4) 0℃抵抗：100 Ω 時，109.6 Ω（when the resistance is 100 Ω at the temperature of 0℃）

12-1) C_j 以外の容量が無視できる場合，時定数を30%まで低減．（Time constant can be reduce to 30% when the capacitance except C_j can be neglected.）

12-2) 9.72°

12-3) 一般に，CCD のほうが有利（CCD has superior noise characteristics in general）

12-4) 入射光の方向に一次回折光が戻る場合（The case the first-order diffracted beam reflects back to the direction of incident beam）

12-5) XPS，FT-IR，分光光度計など（XPS, FT-IR, Spectrophotometer etc.）

13-1) $r_1=r_A/1110$，$r_2=r_A/11100$，$r_3=r_A/99900$　13-2) $r_1=90r_V$，$r_2=9r_V$

13-3) $j20\ \mathrm{k\Omega}$，インダクタンス素子（inductance）　13-4) $B=0.26\ \mathrm{T}$

14-1) $v_{out}=\dfrac{E}{R}\Delta R$　14-2) 0.3 μF　14-3) $d_{33}=2\ \mathrm{pC/N}$，$C_f=10\ \mathrm{nC}$

14-4) $v_o=\dfrac{1}{1+2j}v_i$　14-5) $f_c=\dfrac{1}{2\pi\tau}$

和　文　索　引
Index in Japanese

記　号

ア　行

カ　行

サ 行

タ 行

ナ　行

ハ　行

マ 行

ヤ 行

ラ 行

欧 文 索 引
Index in English

著者略歴

高　偉 Wei Gao
東北大学大学院工学研究科教授
Professor in Graduate School of Engineering,
Tohoku University
Chapters 1, 2, 6, 7, 8, 14

清水裕樹 Yuki Shimizu
東北大学大学院工学研究科准教授
Associate Professor in Graduate School of Engineering,
Tohoku University
Chapters 3, 11, 12

羽根一博 Kazuhiro Hane
東北大学大学院工学研究科教授
Professor in Graduate School of Engineering,
Tohoku University
Chapters 4, 13

祖山　均 Hitoshi Soyama
東北大学大学院工学研究科教授
Professor in Graduate School of Engineering,
Tohoku University
Chapters 5, 10

足立幸志 Koshi Adachi
東北大学大学院工学研究科教授
Professor in Graduate School of Engineering,
Tohoku University
Chapter 9

Bilingual edition
計測工学
Measurement and Instrumentation　　定価はカバーに表示

2017年 3 月 25 日　初版第 1 刷
2022年 8 月 5 日　　第 5 刷

著　者　高　　　偉
　　　　清　水　裕　樹
　　　　羽　根　一　博
　　　　祖　山　　　均
　　　　足　立　幸　志
発行者　朝　倉　誠　造
発行所　株式会社 朝　倉　書　店
東京都新宿区新小川町 6-29
郵便番号　162-8707
電　話　03(3260)0141
FAX　03(3260)0180
http://www.asakura.co.jp

〈検印省略〉

ISBN 978-4-254-20165-9 C 3050